THE NATURE OF THE UNIVERSE

THE WORLD OF SCIENCE LIBRARY
GENERAL EDITOR: ROBIN CLARKE

THE NATURE
OF THE UNIVERSE

Clive Kilmister

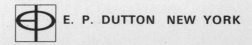 E. P. DUTTON NEW YORK

First published in the U.S.A. 1971 by E. P. Dutton & Co., Inc
Copyright © 1971 by Thames and Hudson Ltd, London
All rights reserved.
First Edition
Printed in the Netherlands.

Library of Congress Catalog Card Number: 76-165333
SBN Number: 0-525-16430-8

CONTENTS

INTRODUCTION

To one of the tribes of ancient India, cosmology was a simple affair. The Earth, like a huge tea tray, was thought to be supported on the backs of three elephants which themselves stood on the shell of a tortoise of universal proportions. The Egyptians likened the heavens to a celestial Nile along which the sun god Ra daily sailed. The cosmologies of both these peoples reflected, as do all cosmologies, what was most important to them in their physical life on the Earth.

So it is in the twentieth century. And because science has become the dominant force in our culture, today's view of the universe is an exclusively scientific one. In this it contrasts even with the picture of the universe drawn for us by perhaps the greatest cosmologist of them all, Sir Isaac Newton. For though Newton's universe was one based on the laws of physics, it was one – as we shall see in this book – which deferred judgement on many cosmological matters to the Almighty. And though today's view of the heavens has perhaps brought us nearer reality through its total rejection of anything other than scientific evidence, it does have one marked drawback. It is both less accessible and less comprehensible to the people that live on this planet than were former cosmologies.

I hope that this book will do something to redress the balance. Cosmology, as the study of order in the universe, is by far the most ambitious of all the sciences. It has inspired the greatest minds and brought about the most profound revolutions in thought yet to have occurred in human evolution. The ebb and flow of ideas in cosmology is constant and the time has long since passed when we can expect any of these ideas to resemble the simple and blinding truth that struck Sir Isaac's head in the orchard.

So in reading this book we must bear in mind that no man has yet been able to explain any natural phenomenon entirely adequately, least of all in the field of cosmology and in that particular form of it known as General Relativity – a theory which, when it was launched on the world, was held to be fully understood by no more than three individuals. In this particular area Professor Kilmister has done a remarkable job. His argument that relativity is in fact no more than well applied common sense is both startling and novel. It supports a private hypothesis I have long held: that no concept in science is too difficult for any enquiring mind to grasp, regardless of the level of scientific education to which that mind may or may not have been subjected.

So I think this book, as well as being an elegant one, is also an important one. Few of us, at one time or another, do not ponder the nature of the universe and the impossibly large distances and concepts which it embraces. But the rise of scientific cosmology has perhaps done something to stifle these primitive philosophical yearnings; we have felt that the matter has been taken out of our hands by the scientist and that our own feelings and views on the matter therefore no longer count (an attitude startlingly similar to that of Newton, who saw sharp boundaries between what he could discover and what was more properly the province of the Deity).

I hope, therefore, that this book will act not as an inhibitor to personal enquiry but as a stimulus to it. Certainly, it enables the reader to see the universe as never before. And if, in this time of multiple crisis, that causes any of us to reflect on the nature of things, I can conceive of no more useful activity for either publisher or author.

Robin Clarke

Probably the most difficult thing of all in cosmology is to conceive the enormous scale on which the universe is constructed. To aid us in the attempt, perhaps we could imagine a book like this one in which the scale represented on each successive page (six inches across) decreases by a factor of 10. To fix our ideas let us suppose that the scale represented on the front cover of the book is 1:1 – in other words, that the cover represents itself exactly. Turning inside, then, page one would represent an area 100 times as large as itself– roughly equal to the size of a very small room. The scale on page two would be 1:100 and the page would accordingly be large enough to contain a drawing of the front elevation of a house. On page three a further scale reduction allows us to plot a fair sized field. You could draw a plan of a small town on page four. Page five is appropriate for drawing the street plan of central London. The width of page six represents a length of about 100 miles, so that the whole of Greater London would fit in very comfortably; while there would be room enough on page seven for a map of the British Isles. Now we begin to approach the astronomical pages, for on page eight we could easily show the whole disc of the 7,900-mile-diameter earth.

From now on the pages begin to get more interesting. If we turn another one, so that the earth now appears as a tiny disc, there is not much to be seen,

Within a tiny area in a galaxy 100,000 light years across, itself one of millions in the universe, is the solar system. Circling the central star at 67,000 miles per hour is the earth (opposite), one of the smallest of the nine planets

Outermost of the planets, Pluto can be seen to move relative to the background of stars. Its orbit is unusual in being very elliptical and at present (until 2009) it is slightly inside the orbit of Neptune, suggesting it may once have been a satellite of Neptune

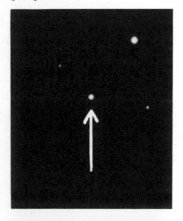

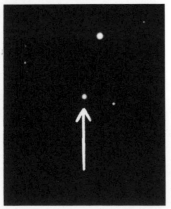

because our nearest neighbour, the moon, is more than 100,000 miles away; in fact, its average distance is about 210,000 miles. On page ten, however, the earth and the moon could be seen as quite a small system, with the earth spinning on its axis every 24 hours, and the moon going round the earth about every 28 days. We turn the page again and the earth–moon system is now seen one-tenth of the size it was before, but nothing else has come into view. It exists there in a large void. It is not until page twelve that we begin to see bodies not unlike the earth in general constitution; they are the other planets. They are moving about in what appears to be a complicated way, but only because the scale is not yet small enough to show their complete paths. Let us leave them for a moment and turn over to page thirteen.

On this page – which has a scale-width of 1,000 million miles – we really begin to see astronomical pictures. We can now see our first star – the sun. This is 93 million miles from us; so we can move the sun to the centre of the page and there is plenty of room for the earth's almost circular solar orbit which takes $365\frac{1}{4}$ days.

The difference between a planet and a star is that a star gives out electromagnetic radiation, most obviously in the form of heat and light, but also as radio waves. The planets are all relatively inactive bodies, like the earth and the moon. They all go round the sun in roughly circular orbits and – significantly, as we shall see – they all lie in the same plane. Mercury, the planet nearest to the sun, takes about 88 days to orbit once, travelling at 30 miles per second. The farthest planet from the sun would not quite get on to page thirteen; but on page fourteen we have all the planets, including Pluto, the most distant, at 3,700 million miles from the sun, orbiting once every 248 years, travelling at three miles per second.

In addition to the named planets there are many small planetary bodies – the asteroids – travelling at various rates round the sun in the same direction as the major planets. We know the exact orbits of about

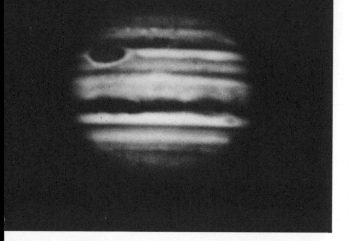

Top: Jupiter, largest planet in the solar system, is nearly 500 million miles from the sun. The coloured belts and Great Red Spot are thought to be caused by the poisonous atmosphere of hydrogen, methane and ammonia. The conspicuous rings of Saturn (centre) are in fact an immense number of tiny particles, probably ice crystals. Its largest satellite, Titan, is bigger than Mercury, and is the only satellite known to have an atmosphere. The orbit of Venus (below left) lies between the Earth and the sun, and so we see it only as a bright crescent at its nearest approach. Dense clouds shroud its surface, preventing close study. Below: the white polar caps and dark areas on the surface of Mars visibly change in extent with the seasons, giving rise to the much disputed conclusion that it might support a low form of life

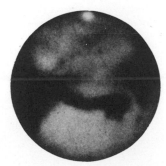

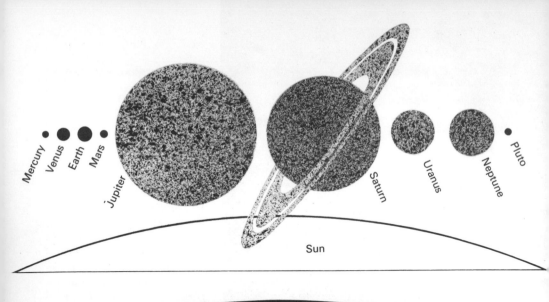

Mercury Venus Earth Mars Jupiter Saturn Uranus Neptune Pluto

Sun

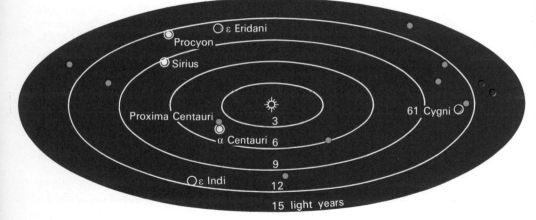

ε Eridani
Procyon
Sirius
Proxima Centauri
α Centauri 6
3
9
ε Indi 12
61 Cygni
15 light years

Top: relative sizes of members of the solar system. Above: extending our view to a distance of 15 light years, we see the sun's nearest neighbours, those ringed being visible to the naked eye. Blue stars have a surface temperature of about 3,000° C, red ones 4,500°C, yellow ones 5,500°C and white ones 10,000°C

1,500 of these rocky lumps, though altogether at least 30,000 have been observed.

'Miles' are now beginning to be an inconvenient unit of distance, as we have so many million of them to deal with. Instead of miles, astronomers prefer to think in terms of the distance which light travels in a certain time; thus the distance between the earth and the sun is said to be eight *light-minutes* and the radius of Pluto's orbit between five and six *light-hours*.

Now to get some idea of the scale of distances beyond our solar system we have to turn several pages beyond page fourteen. On page fifteen we shall see the solar system reduced to one-tenth of its previous size, but nothing else will come into view. Not until page eighteen, when we are considering distances of up to ten *light-years* and the solar system has shrunk to a dot less than one-thousandth of an inch in diameter, do we see any stars in addition to the sun. The nearest, between four and five light-years away, is Proxima Centauri, and at eight and a half light-years there is Sirius. There are also some intermediate faint stars to be seen with good telescopes.

Turning further pages, we can see more and more stars until by page twenty-two we perceive all the stars that are usually visible with the naked eye on a clear, dark night. Among them are all the stars making up the familiar constellations, such as the Plough and Orion. Naked-eye observers around the world can see a total of some 6,000 stars. For every one of these, however, at least 15,000,000 fainter ones are detectable by telescope.

The sun is a rather modest star: its diameter is about a million miles – about four times the distance from the earth to the moon; it is about half a million times as massive as the earth, and its surface temperature is about 6,000 degrees centigrade. We can be sure that the temperature inside must rise very rapidly, otherwise it would not be able to produce the amount of heat and light which it is generating. In order to sustain the necessary thermo-nuclear processes – which

Through the turbulent layer forming the sun's surface streams energy generated by the continuous thermo-nuclear explosion at its centre. Sun spots, regions of relative coolness, appear dark against the glowing gases

A panorama of the Milky Way painstakingly produced by first drawing in 7,000 individual stars with known coordinates and then painting in the more diffuse background from photographs. On the extreme left, below the centre, are the Pleiades cluster and the Andromeda nebula; the dark regions left of centre are obscuring clouds in Cygnus and Aquila. Towards the lower right are the Magellanic Clouds, and above the Milky Way are Sirius and Gemini

resemble what takes place when an H-bomb explodes – the sun's interior temperature must be of the order of 15,000,000 degrees centigrade. The stars in general have surface temperatures ranging from 3,000 to 30,000 degrees centigrade.

Turning to page twenty-three, we find that we can now plot the whole of our local neighbourhood of stars, our galaxy. The shape of the galaxy is rather like two dinner plates placed rim to rim so as to form a circular discus-shaped body. However, because of the rotation of the galaxy, there is an additional bulge in the middle. All the stars that we can see without a large telescope are concentrated within this discus-shaped region. Its diameter is about 100,000 light-years, and its thickness about 3,000. It is rotating slowly about an axis that lies some 35,000 light-years away from us. It takes the sun about 250 million years to go round this central axis. On a clear night, especially in the country, we can see the Milky Way across the middle of the sky – a rather vague band of whiteness as a sort of back-

ground to the main pattern of the stars. The Milky Way is the effect of the millions of stars whose concentrated light we perceive as we look along the plane of our galaxy. Most of these stars are too faint to be individually visible to us with the naked eye.

The stars are not uniformly distributed throughout the galaxy. The formation of the constellations mentioned earlier is partly due to the actual grouping of the stars, but partly, also, to the stars falling in roughly the same line of sight as others that might be much farther away. As well as this spurious grouping, the stars tend to group in globular clusters, each of which may comprise a million stars or more. But for every globular cluster there are probably another million unclustered stars.

In the early years of this century we could have turned no more pages in our book of scales. Apart from speculation by philosophers, everyone would have agreed that when one had described our Milky Way galaxy, one had dealt with the whole universe.

Globular clusters, such as this one in the Centaurus constellation, consist of tens of thousands of relatively tightly packed stars

But let us turn one more page. And now there appears something new and extremely exciting – another galaxy, or nebula, rather like our own. We can actually see it at night in the constellation of Andromeda, faintly visible to the naked eye. It has about the same mass as our own galaxy, is also discus-shaped and similarly comprises many thousands of millions of stars. But by a piece of good fortune, instead of being edge-on to us, so that we see only the rim of the discus, which would tell us very little, we can see that the stars are spread out in spiral arms because of the rotation.

Let us turn one more page, to page twenty-five. Now our galaxy and the nebula in Andromeda are seen as only two out of about seventeen such nebulae forming a cluster known as the Local Group. They are moving hither and thither in a fairly random way, though they have a slight tendency to recede. Some of them are made up of stars; others, perhaps, are made up merely of hot gas, and it is not just a failure in our observation that we cannot detect any individual stars in them. What an astonishing appendix to the story at the beginning of this century these two pages form! But it is even more surprising when we turn one more page – to page twenty-six – for there, about 40 million light-years away, we see the nearest other cluster of nebulae, and this cluster is definitely moving away from us (at about 700 miles per second).

Opposite top: our own galaxy, the Milky Way, as it might appear on page 23 of the book of expanding scales. Sources of bright hydrogen emission are shown in yellow, obscuring clouds in red and young stars and open clusters in blue. Centre: the Local Group, an isolated collection of nebulae surrounding the Milky Way. At the limit of observation with the naked eye – indicated by the red line – is the Great Nebula in Andromeda, M31 (below). Bottom: far beyond the Local Group lie other clusters of galaxies, some large (red), of over 400 members, some medium (yellow) and some small (blue) like ours, of less than 100 members. All are receding from us at hundreds of miles per second

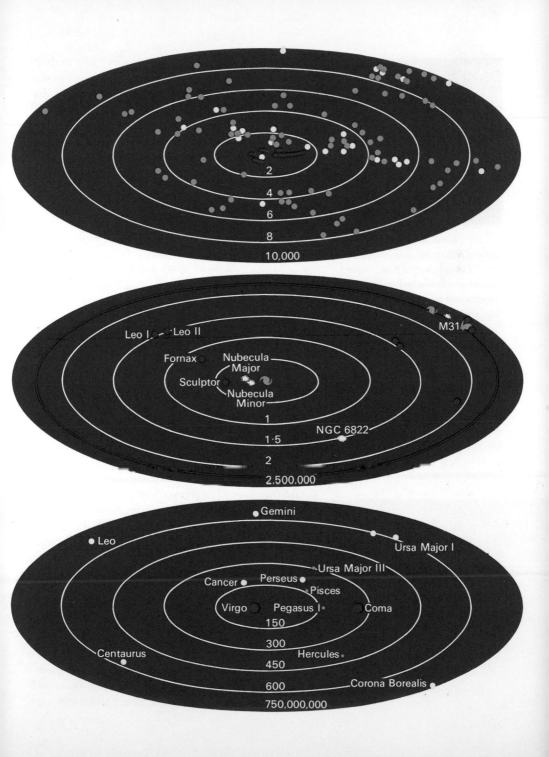

2
4
6
8
10,000

Leo I Leo II
Fornax Nubecula Major
Sculptor Nubecula Minor
M31
1
1·5 NGC 6822
2
2,500,000

Gemini
Leo Ursa Major I
Ursa Major III
Cancer Perseus Pisces
Virgo Pegasus I Coma
150
300
Centaurus Hercules
450
600 Corona Borealis
750,000,000

In 1925 Hubble suggested a classification of nebulae based on three main types. The first ranges from circular (above) to elliptical and includes the brightest and most massive galaxies, born perhaps 400 million years ago. Spirals, the second type, comprise 80 per cent of the catalogued galaxies, and include normal ones, such as our own galaxy and Andromeda, and barred spirals (opposite) whose structure is not well understood. The third type includes the abnormal, shapeless galaxies such as the Large Magellanic Cloud (opposite)

As we turned successive pages, what was perhaps most surprising was the lack of uniformity in the universe. Within the nebulae the stars are not distributed in a uniform manner, but tend to cluster and there is of course the larger-scale clustering of stars to form the galaxies themselves. But even the nebulae are not distributed with any uniformity, but cluster into groups. Our Local Group has some seventeen members, but other groups tend to have many more — anything up to a thousand.

On page twenty-seven we approach the limit of observation with the best telescopes, and at something like 3,500 million light-years away we have the following astonishing picture. Our galaxy, far from being unique or rare, is one of millions of such formations: some, like our own, seen side-on; some spirals like Andromeda; some apparently shaped like spherical or elongated footballs; and others quite irregularly shaped.

So far there is no evidence to suggest that clusters of galaxies are members of greater clusters; rather it looks as if, no matter how far one goes through the universe, one just sees more and more clusters of galaxies distributed in a fairly uniform way. However, this is not the end of our amazement, for as we look we see that the more distant nebulae are all flying away from us. For every extra million light-years' distance they have an additional speed of recession of about 150,000 miles per hour. Thus page twenty-seven takes us near the limit of optical observation, because the light from distant bodies has to pass through the earth's atmosphere to reach our telescopes and when the bodies are so distant the interference with the light caused by the earth's atmosphere is too severe for us to be able to distinguish anything.

So with optical telescopes there is no possibility of turning any further pages and making a chart one-tenth the scale of page twenty-seven – at least not until we place optical telescopes outside the earth's atmosphere. This may soon be practicable on a satellite, but of course it would not be any good putting a

The 200-inch telescope – visible through the open dome – at Mount Palomar is the source of much of our information about the most distant regions of the universe. To reach further into space we need to put such telescopes outside the earth's obstructing atmosphere

relatively small telescope there; we need something of the size of the 200-inch telescope at Mount Palomar, California, to make it a really worthwhile project. Meanwhile, it has turned out that there are other ways of getting information from outer space without light. It was mentioned earlier that the stars, as well as giving out heat and light, give out radio waves. These radio waves from space were first received in 1932 by K. G. Jansky, but he was unable to locate the source of any of them. Such a precise location of source required the vast increase in electronic technology which the Second World War brought about, and, with this, J. H. Hey, S. J. Parsons and J. W. Phillips were able to find very great radio intensity coming from a small patch in the constellation of Cygnus the Swan. This was the first radio source to be pinpointed. Now thousands of radio sources are catalogued.

Very puzzlingly, it seemed at first as if the radio sources were quite independent of the optical ones; almost as if one had a universe of optical stars interpenetrated by an independent universe of radio sources. The first identification of a radio star was by Bolton who located a source in the so-called Crab nebula, which is an object of extreme interest to cosmologists. It is thought to represent the remains of the explosion of a star which was observed by Chinese astronomers in 1054. Of course, since it is some 5,000 light-years away, the explosion must actually have occurred somewhere around 4000 BC. According to Chinese records it was a very striking phenomenon – visible in broad daylight for some weeks and remaining visible to the naked eye at night for several months. It gradually faded, but the nebula seems to be the remains of the exploding material still glowing.

Mystery surrounded the earliest discovered radio sources and there was great argument about whether they were inside or outside our galaxy. There seemed to be a significantly large proportion of them lying in the plane of our galaxy, that is, in the Milky Way; and yet there were a number that were not. This question was finally settled when it was realized that there were two families of radio sources, some of which were indeed galactic and others extra-galactic. The mystery about the extra-galactic sources then deepened further, because it seemed that they could not be so far away and yet be such strong sources. Nonetheless, they were. The first radio source of all, the one in Cygnus, was accurately pinpointed by Cambridge astronomers, and observers at the Mount Wilson Observatory, California, were able to correlate it with a pair of

The second strongest radio source in the sky is Cygnus A (above), a strange double object in the constellation of Cygnus. The visible galaxy was found to be lying midway between two areas of radio emission (below), the contour lines indicating intensity increasing towards the centres

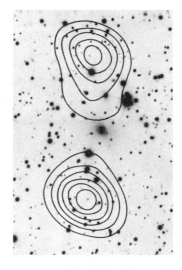

optical nebulae. It was then thought that the nebulae were colliding and that the radio waves arose from the complicated and violent physical circumstances of the collision. Much time was wasted searching for colliding galaxies, and a number of other radio stars were identified with single galaxies.

The enormous amount of energy that seems to be coming from some of these sources remains a mystery to us. This has become even more perplexing in recent years, because among these sources there are a few which are extremely small. And some of these sources are variable; that is to say, the light from them, or rather the radio emission, increases and decreases at different times according to a definite law. It is hardly possible that an enormously large object like a large galaxy could be variable in this way. In fact, because of the variability, the size of these objects must be something comparable with that of a star or even smaller. And yet the amount of energy produced is far in excess of the amount which could be produced by thermo-nuclear methods in material the size of a star. We are obliged to find some other source of energy that could keep these radio beacons going. These objects are the quasars, about which so much argument has raged in the past few years.

In this way the radio telescope has extended the range of our knowledge to cover one more page in our imaginary book. And now we really do seem to be near the end. On page twenty-eight we find distant objects, some of them receding from us with velocity comparable to the enormous speed of light, and, as we shall see, this is an indication that we cannot expect by any observational means to probe very much farther.

The time scale

We have now dealt with the spatial distribution of matter in the universe. It remains to say a few words about the time-scale involved. Archbishop Ussher (1581–1656), on the basis of the Old Testament, estimated the creation of the universe to have taken place in 4004 BC. In more recent years historians have

dated the building of the ancient city of Ur from at least 8,000 years ago, so that the earth must have been in very much the same state then as now. Rather more emphatically, however, the evidence of modern geology shows that at least 4,000 million years were necessary to bring the earth to its present geological state. In fact, the cosmologists believe that the actual age of the earth is something of the order of 4,000 or 5,000 million years. Among the stars in our galaxy the general age seems to be something of the order of 6,000 million years and this estimate has recently shown a tendency to rise.

Of course, our observations tell us nothing about what we might see if we were able to transport ourselves back in time to a point, say, 4,000 million years ago. The universe at present is expanding. Many people feel that if time could be reversed, like a cinema film wound backwards, one would gradually find the matter returning to a more and more concentrated form, implying that the present expansion is the continuation of an original explosion. But this simple-minded way of looking at things, attractive as it may be in appearing to make no assumptions beyond what we see, is in fact a very hazardous generalization indeed. We shall show in later chapters that it may be more reasonable to take other views about what we might have seen all that long time ago.

When was the universe created? Archbishop Ussher dated the event as recently as 4004 BC, but even the ancient Sumerian city of Ur (above) was standing at least 8,000 years ago. Modern estimates of the age of the universe are in the region of six thousand million years

SPACE AND TIME
FROM THE GREEKS TO NEWTON

The last chapter has served to give a brief indication of the extent of our present-day knowledge in cosmology. Let us now go back and, instead of taking for granted everything that scientists tell us, try and trace the sources of this knowledge.

Wandering stars and old philosophies

To look up into the night sky must have been, for primitive man, a very moving experience. In those days, long before the devastations of industry, a cloudless night was a very clear night and to go out into the open was to see an amazing picture of brilliance. Stars, or points of light of all kinds, made persuasive patterns in the sky, into which man could read pictures of animals, battles or anything that he wished. Moreover, on going out at about the same time on successive nights, he would begin to notice that these patterns changed, and that the whole picture was slowly rotating. Indeed, the motion of the stars must have led early man to the first severe philosophical problem which besets the subject of cosmology, the problem of change.

How is it possible for something to change without losing its identity? The pattern of stars on the sky appeared to remain the same although being in a different position, and some idea of the motion of the earth and the stars must have arisen from this observa-

Opposite: from earliest prehistory man has seen visions of battles, ships and animals in the night sky and used the celestial bodies as clock, compass, calendar and god. The signs of the zodiac – the narrow section of the sky along which the sun seems to move – are still used today to identify the brighter constellations

Thales' cosmology of a flat earth floating on water. The central figure is Archimedes, and around him are the other three elements – earth, air and fire. Across the heavens move the signs of the zodiac

tion. The Greeks considered these problems in connection with their theories of the universe. Even before Socrates' time there was the beautiful idea of Thales (640–550 BC) that the earth floated on water like a ship – tossing on the waves (which accounted for earthquakes). But Anaximander (611–545 BC) altered this to a much more beautiful scientific theory, in which the earth was regarded as a drum held stationary in space by the equal forces of the equidistant bodies surrounding it. This revolutionary idea made possible the later cosmology of Copernicus in which it was recognized that the earth moved round a fixed sun. Anaximander's construction, because it was so elaborate, led him also to the problem of change.

Our upbringing from childhood is moulded by the philosophical framework which is current today, so that we find it hard to understand the magnitude of this problem for the pre-Socratics. Yet Parmenides, a most subtle thinker, became convinced of the non-existence of change, and so concluded that change was only an appearance and that motion (a change of position) was impossible. This conclusion is evidently contradicted by the facts, and it was this situation that was tackled by Democritus (460–370 BC). Beginning with the fact of motion, he argued that the world must have parts, and that (contrary to Parmenides' view) it cannot be full. Since it is not full, there is a void between the parts. The parts themselves are unchanging – here Democritus is taking Parmenides' view, only applying it to the individual 'building bricks'. Change is due merely to the re-arrangement of the parts or 'atoms' in the void – a theory which held sway till the early years of this century. Democritus mostly interpreted the atoms as being invisibly small, as he had to do for the sake of agreement with observation, and so it is not fanciful to see him as the first person to connect changes in the sky with local phenomena, and so to see a unity between matter on the very large and the very small scale, a unity which is still for the most part a mystery to us today.

Let us return now to the pattern of the stars. It was a natural hypothesis, although it later proved inconvenient, that the general motion of the stars is caused by their being fixed to a large, slowly rotating sphere which is centred on the earth. This explains why the whole pattern is swept past one's vision, but one of the inconvenient things in this theory is that not all the bright points of light which one sees on a dark night move in the same direction.

The system described here, with a sphere for the stars, is the Ptolemaic system in its simplest form. The name is misleading; it was the ultimate form of Greek astronomy, but it is through Ptolemy's *Almagest* (about AD 150) that it has come down to us. There was an interior sphere for man, and an exterior sphere for the

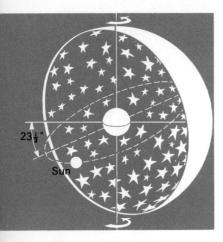

23½°

Sun

The simple, earth-centred system of the early Greek philosophers, with the sun moving along the ecliptic and the universe bounded by a rotating sphere of stars

A sixteenth-century diagram explaining the phases of the moon according to how much of the illuminated surface is visible from the earth (the central eye)

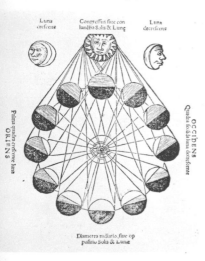

stars. In between, there was the sun and a void. Outside the sphere on which the stars were fixed there was nothing. The sphere of the stars rotated steadily once every 23 hours 56 minutes on a fixed axis. The sun, on the other hand, in order to fit in, turned round on its axis and also moved eastward on its track, the so-called ecliptic, tilted at 23½ degrees to the equator, once in 365¼ days.

The different combinations of stars that people in different latitudes on the earth might expect to see were satisfactorily explained by this arrangement, and so were the days and the nights, the changing seasons and the nightly motion of the stars. Indeed, this model is so satisfactory that it is still used to teach navigation and surveying. The planets present a problem because they are the points of light whose movements are not explained by the model. They are bodies having much the same appearance to the naked eye as the stars, but with a more complicated motion. In the first place, they obviously follow roughly the general westward motion of the stars, but, on the other hand, they have an additional eastward motion relative to the stars, which leads them to make a complete revolution and so to return more or less to their original positions in relation to the stars. This can be explained with the two-sphere theory, for the planets can be imagined to move in the void between earth and stars. But matters get worse.

The moon, which was naturally counted by primitive man among these bodies, travels around the ecliptic faster, and also less steadily, than does the sun. And the moon's phases, of course, are the oldest of all the calendar units, because they are easily visible. But the simple lunar unit soon turns out to be a hopeless basis for a calendar, because successive new moons may be separated by intervals of either 29 or 30 days. Indeed, it was many years before the length of a particular specified month could be satisfactorily determined by mathematical theory.

The periods of complete revolutions of the planets – Mercury, Venus, Mars, Jupiter and Saturn – are differ-

The heavenly spheres according to Aristotle, as understood in the Middle Ages. Around the earth moved the sun and moon and the five other planets then known. Beyond was the firmament of stars, the sphere of the prime mover and heaven

ent from each other; furthermore, a single particular orbit may be quicker than the average period of the planet; and, worst of all, the planets' movement, generally eastward relative to the stars, is periodically interrupted by periods of 'retrograde' westward motion. Moreover, during these periods a planet may increase in brilliance. For example, Mars, when in opposition to the sun, is brighter than anything in the night sky except the moon and Venus.

To explain the various planetary motions the Greeks proposed that there was a series of spheres moving as though interlocked. This theory was due to Plato's pupil Eudoxus, and was a central feature of Aristotle's

Ptolemy explained the retrograde motion and varying brightness of the planets by the mechanism shown below, by which all planets moved round the earth in a series of loops, or epicycles

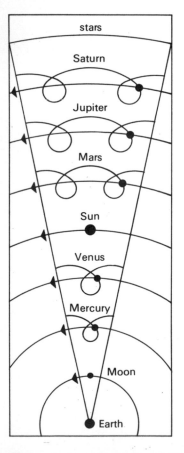

world-picture which was the next, more general, version of the Ptolemaic system. Aristotle must be counted as the most comprehensive and influential cosmologist of the ancient world. But no matter how ingeniously such a model was elaborated, it was quite impossible for it to account for the variation in brilliance during retrograde motion, or other phenomena revealed by improved techniques in observation.

The next step forward was the proposal by the two Greeks, Apollonius and Hipparchus, of the famous 'epicycle' mechanism. The simplest form of this mechanism is the spirograph, in which a small circle, the epicycle, rotates uniformly about its centre, which is a fixed point on the circumference of a second rotating circle. In the spirograph the outer circle rolls on an inner one, but this is merely because it is more convenient. The planet is on the epicycle, and the centre of the other circle is the earth. In this way a path is described for the planet which accounts for the retrograde motion and for the increased brightness of the planet (because it gets nearer to the earth), so that many of the irregularities can be explained. But of course, in order to describe all the planets' motions, a separate epicycle system has to be designed for each one. The loops vary in size, the diagram for Jupiter requiring eleven loops for a complete revolution and the Saturn diagram requiring 28. By using these variations, the system of circular motions could be made to describe pretty accurately everything about the planetary motions. The epicycle system showed itself capable of astonishing generalization, and here Ptolemy was the main contributor. One could add a small epicycle whose centre moved on the original epicycle whose centre, in turn, continued to move on the circle centred on the earth. In this way, and with other tricks, a surprising amount of the planetary motions could be explained under the system.

In some ways the epicycle theory is a wonderful scientific theory; in other ways, it is a nightmare. After all, the scheme is intended as a descriptive one – that is, a device to make it easier to remember and predict the

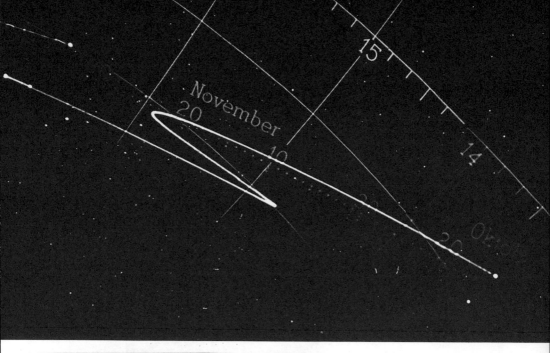

Above: a small section of the actual path of Mars, illustrating its retrograde motion. Left: Ptolemy, measuring the altitude of the moon, guided by the muse of astronomy

The revolutionary work by Coperni-cus on the movement of the planets lay hidden for some 40 years before its publication in 1543, in the face of severe opposition from the Church and followers of Ptolemy. It re-affirmed the theory of Aristarchus, 300 BC, that the sun, not the earth, lay at the centre of the universe

motions. One begins to wonder whether the motions are not more simple when they are considered just as they are, rather than with such a complicated mechanism to explain them. Nonetheless, for centuries this cosmological theory of the Greeks prevailed, because there was nothing to replace it. The first real challenge was that made by Nicholas Copernicus in the sixteenth century.

Copernicus was a 'universal man'; he studied languages, law, theology and medicine, as well as mathematics and astronomy. His studies soon convinced him that the Ptolemaic system would not do. There were too many significant discrepancies between theory and observation. Moreover, he had found among the classical works several suggestions quite at

variance with the Ptolemaic system which he wanted to investigate. As long ago as Aristarchus, it had been suggested that the sun, and not the earth, was in the centre of the universe so that the sphere of stars had its centre on the sun. The earth, notwithstanding ordinary man's experience, rotated about an axis, thus producing what was only an illusion that the heavens rotated in the opposite direction. By placing the sun at the centre of the planetary system as well as that of the stars, and assuming that the planetary orbits were circles, Copernicus was soon able to satisfy himself that the planetary loops and the change in intensity of light could be explained without the complicated machinery of epicycles.

Indeed, Copernicus was able to go further and show that the distances of the planets from the sun could be found in terms of the radius of the earth's orbit, the so-called astronomical unit of distance. At a time when all knowledge had to be sanctioned by the Church, Copernicus' work had a fairly lenient passage. But the pendulum gradually began to swing back and by the time Galileo had used the newly invented telescope to verify Copernicus' assumptions, the Church's attitude was much less liberal and Galileo was tried before the Inquisition. This religious intervention brings us to the end of a chapter of man's history.

An advance in causal theories

We must draw a careful distinction in science between a purely descriptive theory and one which attributes 'causes' to events. Among descriptive theories, perhaps the best-known examples now are nineteenth-century zoology and botany. The animals and plants were classified into species but there was no theory to explain why any of these species should have taken this particular form rather than any other. A classificatory science requires immense effort but does not produce the kind of explanation which we seek in astronomy or physics in general, and which we need especially in cosmology. Both the Ptolemaic system and that of Copernicus were purely descriptive systems.

Aided by his famous telescopes (below) Galileo discovered Jupiter's four moons (shown in the 1660 drawing opposite) and the phases of Venus. In 1663 he was forced by the Inquisition to deny his agreement with Copernican theory

Building on the years of systematic observation by Tycho Brahe – shown below with his mural quadrant at Uraniborg – Kepler attempted to illustrate the harmony of the universe in terms of the size of the planetary spheres, related to successive Platonic solids (right). His observations in fact led to the formulation of his three laws (opposite): planets move in ellipses, with the sun at one focus, A being a geometrical point only (top); equal areas are swept out in equal times by a line connecting sun and planet (centre); the cubes of the average distances of the planets from the sun are proportional to the squares of their periods – illustrated here for Earth and Mercury

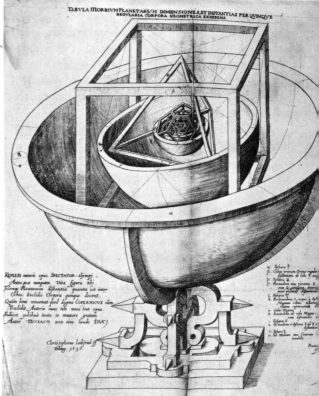

The notion of cause is a very old one, at least for events on the earth's surface, and the idea of pushing a barrow, or shooting an arrow, shows the application of this concept to motion. The next question is whether there is some causal theory which will explain the planetary motions, and this brings us on to Johannes Kepler and Isaac Newton. Kepler's work is in many ways a bridge between the earlier descriptive theories and the causal theories, because he was seeking only to describe the observations and so produce a descriptive theory, and he did this very successfully. Consequently the laws of planetary motion that he managed to deduce were of great value to Newton in finding his causal theory.

Kepler, after a lifetime's careful observation, often motivated by curious metaphysical theories, formulated the following descriptive laws of planetary motion:

1. That the planets move not in circles, but in ellipses, one focus of each ellipse being the sun. The focus of an ellipse is one of the two points which, in the case of the gardener's construction for marking out elliptical flowerbeds, would represent the two stakes which, when connected by a slack string, may be used to describe the ellipse.

2. The radius from the sun to the planet sweeps out equal areas in equal times.

3. The cubes of the average distances of the planets from the sun are proportional to the squares of their periodic times.

These three laws, although still purely descriptive, are an immense advance over the epicyclic theory. Perhaps their most important advantage is that they separate the different mechanical features that need explanation. The first law says nothing about how fast the orbit is described, but states only something about its shape. The second law then states something which is equivalent to how fast the planet moves at one point of its orbit compared with another. The remaining law relates the speed of one planet to another. The essential method of science is exemplified here: to break down complicated phenomena into simple units which can then be explained individually.

Now for hundreds of years, progress in mechanics had been held up by the idea that a force of some kind was needed to produce velocity. Newton, in trying to understand Kepler's laws, realized that it was not right to think that a force is needed to maintain velocity, notwithstanding the experience of earth-bound man in trying to push a barrow against the force of friction at the wheels, or the evidence provided by a block of stone sliding on the rough ground. Instead, Newton took the view (following the more limited idea of Galileo) that a force is not needed to produce velocity but that it is needed to produce a change of velocity,

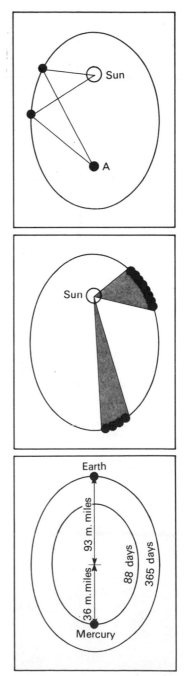

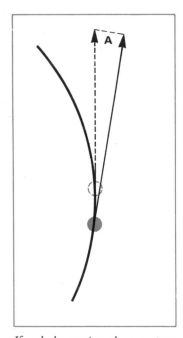

If a body moving along a curve suddenly found itself under no forces, it would continue along a tangent. Newton (below) deduced from the fact that the velocity changed (A), even if only in direction, that a force must be acting in the direction of the change

or an acceleration. His first law of motion states that a body under no forces moves with uniform speed in a straight line. Force is therefore introduced as that which causes a change in such motion, and hence a change in the velocity.

What relevance have these laws to Kepler's observations on planetary motion? To understand this better we have to realize that acceleration is produced not only when a velocity changes in magnitude but also when it changes in direction. And a planet travelling along an elliptical course is changing its direction of motion all the time.

Consider, for example, a heavy object being whirled round in a circle at the end of a length of string. Newton's laws tell us that if the object were left to its own devices it would fly off at a tangent. Instead, the string is constantly pulling the object towards the centre of the circle. There is, as we all know, a tension in the string, acting as an outward force on the hand that holds it and as an inward force on the object. This latter force is, in effect, making the object accelerate constantly towards the centre of the circle.

Now the planets are not tied to the sun by pieces of string and Newton reasoned that some similar force must be acting on the planets to keep them in their elliptical orbits. If there were no such force, the planets would simply fly off in a straight line into the depths of space. This force Newton called gravitation.

Thus Newton was able to show that Kepler's laws were comprehensible if there was a gravitational pull between the sun and planets. By further generalization he inferred that there was an attractive gravitational force between any two bodies. Newton took the acceleration produced in a body as a measure of the force acting on it. But, of course, everyday observation shows us that some bodies are moved more easily than others, hence each body must have some quantitative assessment, and this he called the 'mass'. It is a measurement of the body's resistance to change of motion. The mass, for Newton, was a constant, attached to the body once and for all and interpreted as the 'quantity of

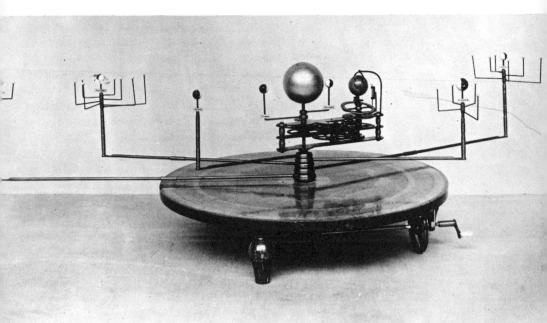

matter' in the body, which measured its inertia. The force was then measured by the product of the mass and the acceleration: this is Newton's second law of motion. With his first two laws of motion, and the third, which states that the action of one body on another is equal and opposite to the action of the second on the first, and takes place along the line joining them, Newton was able to deduce all three of Kepler's laws of planetary motion, thus carrying over this descriptive system into a perfect causal description of the motion of the planets. When he had done this, it became clear that the consequence of Kepler's second law was that the planets moved under a force towards the sun; the first law then implied that this force varied inversely as the square of the distance, so that if a planet is twice as far from the sun, the force on it is divided by four. The third law then implies that the force between sun and planet is proportional to both their masses.

Newton's law of gravitation had widespread implications, which have now become so engrained in our thought that we tend to take them for granted. At the

Orreries – mechanical models of the solar system – became popular in the eighteenth century as a result of the interest in astronomy stimulated by Newton's theories. Uranus, Neptune and most of the satellites shown here were unknown to Newton, but they were deduced through his law of universal gravitation

time, it was very surprising to learn that what made the apple fall to the ground was the same force that kept planets in their orbits round the sun. In other words, Newton clearly believed that this law applied not only on the earth's surface but throughout the planetary system. There is some evidence that he also thought the law would apply throughout the universe. Thus was born the idea that the laws of physics could be applied to any cosmological problem – an idea which turned cosmology away from metaphysics and brought it within the embrace of natural science.

But we should examine in more detail the assumption that Newton's laws are valid in all places and at all times. There is one everyday experience which seems to indicate an exception. Travelling upwards in a fast lift, one has the sensation that one's weight is increasing, and a briefcase feels heavier than it did when the lift was stationary. This is not because the lift is simply moving fast, but because it is accelerating. Similarly, as an aircraft speeds up for take-off, objects such as books on smooth surfaces tend to slither backwards, an apparent contradiction of Newton's law, since this law states that a force is needed to produce their backward acceleration.

The lift and the aircraft are frames of reference in which Newton's laws do not apply, because they are accelerating frames of reference. Before the laws can apply again, a 'transformation' must be made. That is, the superimposed external acceleration must be subtracted from all the bodies involved before the laws are applied. The weight of the briefcase then stays constant and the books in the aircraft in effect stay in the same place.

In Newton's view there is a crucial difference between acceleration and velocity. Acceleration, Newton argued, is an absolute thing and there are frames of reference which are not accelerating. Under such conditions, objects do not move faster or slower unless a force acts on them – they continue to move with uniform velocity. Thus there is no need to speak of relative accelerations; instead, all accelerations can be measured against the condition of no acceleration.

In the case of velocity the situation is quite different. It is impossible to define such a thing as absolute rest, since there is no standard against which to measure it. A train moves relative to the earth, the earth moves relative to the sun, the sun moves relative to the Milky Way, and the galaxies move relative to other galaxies. There is no absolute standard. One can speak only of things moving with velocities relative to other things.

Newton argued that his laws applied in all these reference frames, provided that they were not accelerating. These frames are called inertial reference frames in recognition of the fact that an object's inertia makes it remain at rest or in uniform motion until it is acted on by an external force. When a car goes round a bend, its passengers tend to continue straight ahead. Only by holding on, and exerting a force against their seats, do they stay with the car's motion.

This means that a game of billiards, for instance, will proceed according to the same rules of mechanics on a train moving at constant speed as in a signal box at rest relative to the earth's surface (always assuming the train ride is perfectly smooth). Moreover, the signalman can work out how the billiard balls will behave on impact with one another on the train, even though he himself has a different relative velocity to the train. To do this, however, he must first subtract the train's velocity from the speed of the billiard balls. He can then work out the effects of the impact according to Newton's laws. But to produce a result as it would be seen by a passenger on the train, he must then add on the train's velocity to the motions of the balls after impact.

Mathematically, this is very simple to do and involves only the idea of transformation which I have already mentioned. The signalman transforms the problem from the train to the signal box, works out the solution, and then transforms it back to the train again. This might seem a pointless exercise, but in fact it is not, for some problems are very much easier to solve from the signalman's point of view than from

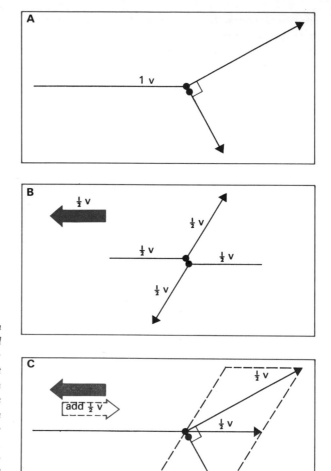

In a game of billiards a passenger on a train might see a stationary ball hit by a moving one with speed 1v and the two balls then move off at right angles (A). A signalman with respect to whom the train happens to be moving with speed ½v sees a more symmetrical impact (B), the result of which seems obvious to him. He can then calculate what the passenger sees by adding the speed of the train (½v) to the apparent velocity of the balls (C)

the passenger's. An even more vivid illustration of this will be given with the description of how Einstein transformed gravity out of the universe, and so arrived at his space-time continuum.

Any theory of universal gravitation must be, in effect, a theory of the whole universe. Nearly all present-day cosmologies have been developed first from Newton's ideas and more directly from Einstein's modifications of them. Newton himself, however, did not pursue the startling implications of his theory as far as might have been expected.

He did reason, however, that the universe must extend indefinitely in all directions. Aristotle had postulated a universe of finite size and most medieval thinkers agreed with him, but Newton's discoveries ruled this theory out of court. For if the universe were finite, he said, the gravitational force would attract all bodies towards each other and the universe would end up as one massive lump of matter. Only if some of the matter were infinitely far away, so that its gravitational attraction towards nearer pieces of matter were infinitely small, could this question be satisfactorily resolved.

In fact, there was a paradox in this argument which was not to be pointed out until nearly the end of the nineteenth century. It turns out that if matter is evenly distributed within a sphere of certain radius, the gravitational field on the edge of the sphere will be proportional to the radius. It follows that the gravitational field at the edge of an infinite sphere – which is what is meant by an infinite universe – will be infinitely large. Just what chaotic physical conditions infinite forces could lead to was not something physicists cared to reckon with, and this led the German astronomer, H. Seeliger, to propose in 1895 that Newton's law of gravitation must be modified if very large distances were being considered.

Newton himself realized that his theory of an infinite universe assumed that his laws applied throughout the universe. Whether he also fully realized the difficulties this might lead him into is not clear. He was, however, wise enough to leave the matter entirely open, and towards the end of his famous work, *Opticks*, he deferred final judgment to the Almighty: '. . . it may also be allow'd that God is able to create Particles of Matter of several sizes, and in several Proportions to Space, and perhaps of different Densities and Forces, and thereby to vary the Laws of Nature, and make Worlds of several sorts in several parts of the Universe. At least, I see nothing of contradiction in all this.'

Fig. 1.

TO GEORGE THE THIRD KING OF GREAT BRITAIN &

This View of a Forty Feet Telescope, constructed under his Royal Pa

h permission, most humbly inscribed, by his Majesty's very devoted and Loyal

and most grateful obedient Servant, William Herschel.

Once Newton's laws of motion had been proposed and their consequences worked out, the tremendous triumph of his achievement was realized. A very full description of the planetary motions was provided; indeed, those planets which did not move in ellipses as the theory predicted were found to be moving under the influence not only of the sun but also of the remaining planets. When the influence of these remaining planets was taken into account, the correct motion of the planets was calculated in almost every case and the success of this method led to the discovery of new planets. Until 1781, Saturn, known from the earliest times, was still the most remote planet known; then Sir William Herschel found the planet Uranus during a routine survey of the sky. It has since been realized that Uranus was observed about twenty times between 1690 and the date of discovery, but in each case had been regarded as an ordinary star. Herschel's observations enabled him to determine the details of the planet's orbit, but these details could not be confirmed by later observation. By 1820 the discrepancy was considerable and it was realized that the planet's position could not possibly be reconciled with its earlier position or with Newton's Law of Gravitation.

The Law of Gravitation was suspected by some people, but John Adams, then an undergraduate in Cambridge, and Urbain Leverrier, a young French

William Herschel discovered the Georgian Planet (Uranus) and two of its five moons in 1781, using a telescope he had constructed himself. His colossal 40-foot telescope (opposite) and more accurate 20-foot instrument enabled him to study and later compile the first extensive catalogue of nebulae

It was as a result of calculations by Leverrier (below), based on the irregular motion of Uranus, that Neptune was first observed in 1846. At this time only one satellite (arrowed) was seen. Adams (bottom) had predicted the planet's existence a year before, quite independently, but his results were neglected

astronomer, both worked on the basis that the Law of Gravitation was correct, and concluded, by 1846, that another planet beyond the orbit of Uranus was distracting it from its predicted orbit. The Newtonian theory enabled Adams and Leverrier, using different methods, to solve the problem of finding the position of the new planet in a purely mathematical way; and when in 1846 a telescope search from Berlin was started, the planet was found almost exactly in the predicted position. This planet was Neptune and a similar situation has since led to the discovery of the farthest known planet, Pluto.

There were, however, further complications towards the centre of the solar system. The orbit of the nearest planet to the sun, Mercury, rotates slowly and this rotation is to a large extent accounted for by the gravitational forces of the other planets. But there is a small residual rotation which does not seem to be accounted for. Here again the Law of Gravitation was at first suspected, although other astronomers were led to calculate the position which a small planet, called in anticipation Vulcan, would have to occupy to produce the changes in the orbit of Mercury. Unfortunately, the telescope has never been able to provide any evidence of the existence of Vulcan, and astronomers nowadays are not disposed to accept the existence of bodies on the strength of a calculation alone. In fact, an alternative explanation has been found for the behaviour of Mercury and this will be discussed in a later chapter under the General Theory of Relativity.

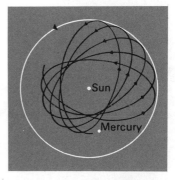

Thirteen times each century, Mercury passes between the earth and the sun, appearing as a tiny black dot against the sun's disc – evidence that it has no atmosphere. It was thought that tidal forces kept the same face of Mercury towards the sun, like the earth-moon system, but recent radar measurements show that it rotates once each complete orbit. The orbit is really a rotating ellipse, exaggerated in the drawing above

45

But so successful was the Newtonian theory that many questions that needed to be asked about it were not asked until very much later. The first question, which in many ways was the first genuine cosmological question, was this: when one predicts and observes that the orbit of a planet is in the form of a rotating ellipse, what is one's standard of 'fixity' or non-rotation? Not, of course, the surface of the earth, for, as we know, the earth is rotating on its axis and astronomers, in order not to lose sight of a particular star, have to counteract the rotation of the earth by moving their telescopes in a systematic way. The observations and the theoretical predictions have been made in a frame of reference in which the distant stars do not, on average, have any transverse motion. Fifty years ago one would have said 'in which the distant stars are fixed', but we have learnt to be a little more careful now, because they can appear to be fixed but in fact be moving directly away from us. The important thing is that there should be no systematic rotation of the frame of reference relative to the distant stars.

Because these stars had been used for hundreds of years as a standard by which to observe the planetary orbits, it was originally not surprising that they defined the frame of reference. But from a theoretical point of view it is very surprising indeed, and evidently an extremely significant point about the actual universe. The Newtonian predictions are carried out relative to the inertial frame mentioned earlier. These frames of reference are the ones in which Newton's laws are valid and so can be determined by experimental devices, without any apparent relation to the fixed stars. Jean Foucault's pendulum, first demonstrated in 1851, is such a device in which a bob is suspended by a long thread from a point so as to swing freely to and fro. If it is allowed to move without initial rotation, easily achieved by tying the bob to one side by a thread and burning through the thread, the plane in which the pendulum swings will be found to rotate slowly relative to the earth's surface. If we keep our telescope fixed in such a plane rotating with the pendulum, we

Opposite: Foucault demonstrated the earth's rotation by his famous pendulum experiment in the Panthéon in Paris in 1851. As the earth rotated about its axis the plane of oscillation of the pendulum changed with respect to the scale fixed to the floor

shall find that not only have we successfully counter-balanced the effect of the earth's rotation, but we also have a frame of reference in which Newton's laws apply accurately.

No mention has been made here of the so-called 'fixed stars', and yet we have determined a frame of reference in which these distant stars are fixed. There must, then, be some connection between the local inertial frames of reference and distant matter. Since we can hardly suppose that the choice of the frames of reference is influencing the distant matter, we are forced to take the other view, originated by the German physicist Ernst Mach in 1893, that the local frames of reference are determined by the distribution and possibly the motion of the distant stars.

This conclusion is nothing short of amazing. How could the distant stars determine, in effect, how the laws of physics worked on the earth's surface? To understand this better, we must go back to the astronomical observations made in the eighteenth and nineteenth centuries.

Beyond our solar system

When we look with the naked eye at the stars, they are not uniformly distributed throughout the heavens, but form a broad beam across the middle, the Milky Way, with a relatively small number in other directions. Of course, it was not always known that the Milky Way was made of stars, and discussions about its nature occupied many thinkers. Astronomers generally paid little attention to their conclusions until the middle of the thirteenth century.

Perhaps the most outstanding of these earlier contributions to cosmology is that made by Immanuel Kant in his *Theory of the Heavens* (1755), his first important book. There can be no doubt that the cosmology of Copernicus and Newton was the most remarkable inspiration for Kant's contributions. He formulated what was later known as the Kant-Laplace hypothesis on the origin of the solar system: that it is formed from an enormous rotating cloud of gas which threw off

Opposite: the North America nebula in Cygnus. To early astronomers, the visible stars in the Milky Way appeared to be surrounded by pale nebulous clouds. With the development of increasingly powerful telescopes we have been able to resolve the thousands of stars present in these areas

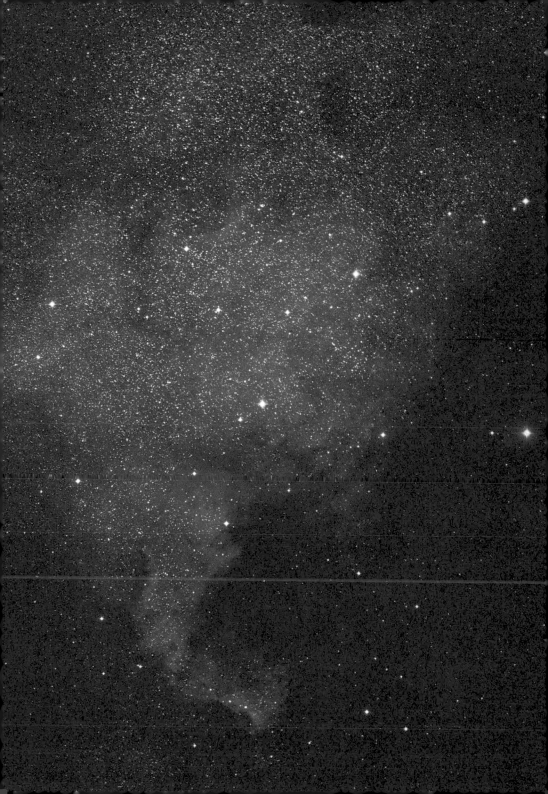

 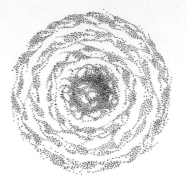 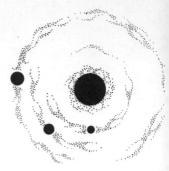

Kant believed that the solar system began as a giant cloud of gas which, as it contracted gravitationally, started to rotate. Rings of hot gas flung from the centre later condensed to form the planets

the planets (thus explaining why they are all in one plane), and the remains of which became the sun. But he also applied the same idea to the Milky Way, which had been interpreted as a system of stars by Thomas Wright five years earlier. Not only this, but Kant was led to surmise that the most distant nebulae, which had earlier been thought to be stars, might indeed be clusters of stars of the same kind as our own Milky Way. Kant was not only astonishingly ahead of his time in this matter, but it is worth mentioning that the cosmological problem – that is to say, the question of whether the universe was finite or infinite, both in space and time – was what led him to his theory of knowledge and so to his *Critique of Pure Reason*.

Ideas like Kant's were somewhat frowned upon by astronomers until Sir William Herschel set out to map the Milky Way systematically. He did this on the basis of certain assumptions which unfortunately have little foundation in fact. He began making a catalogue of all the double stars he could find – that is, pairs of stars each of which is revolving in the gravitational field of the other. He hoped to use information from this in making measurements of distances. At the same time he made measurements of relative brightness. Here he assumed that the intrinsic brightness of all stars is much the same, so that the distance of a star is proportional to its magnitude or faintness. The astronomer's scale of magnitude increases with faintness,

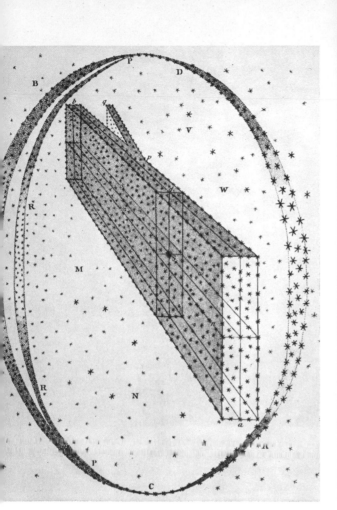

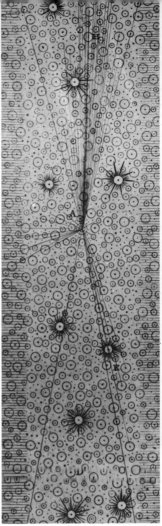

implying that the less bright the star is, the further away it is. There was already considerable evidence that this was not the case. For example, the stars in the Pleiades are distinctly different in brightness although they are all roughly the same distance from the earth. He assumed also, in making his chart of the Milky Way, that the stars were uniformly distributed in space and that he could penetrate with his telescope to the outermost regions of the Milky Way. In this way, counts of the number of stars that were visible in a particular direction gave him some idea of how far the Milky Way extended in that direction.

Above, right: Thomas Wright's diagram of the Milky Way. Seen from A, the stars vary in magnitude, but appear tightly packed despite the great distances separating them. Left: Herschel's diagram, published in 1784. An observer on earth (S) sees the Milky Way (ab) as a great encompassing belt of stars (ABCD), varying in density with direction of view. A branch (pq) of the Milky Way would appear as a narrow luminous strip (PRRP)

The Pleiades cluster and nebulosity in Taurus, clearly showing the difference in brightness of the individual stars

Later on, however, when Herschel had a larger telescope, he found that there was no limit to the distribution of the stars, and he also discovered much about the nature of the nebulae.

He had been puzzled and fascinated by the nebulae for most of his life. These had by now been catalogued by Charles Messier (1730–1817), and Herschel used his large telescope to try and test Messier's guess that most of the nebulae were composed of individual stars. In 1790 he found a nebula in which there was a central star, and the rest of the nebula appeared to be merely a gaseous atmosphere surrounding it. This shook his confidence in the validity of the hypothesis. It need not have done, of course, for the explanation was that the telescopes necessary to resolve the nebula had not been constructed.

A considerable advance was made by an American, Cleveland Abbe, in 1867. In his paper based on a study of Herschel's *Catalogue of Nebulae and Clusters of Stars,* he proposed (quite correctly as we now realize) firstly that the ordinary clusters of stars belonged to the Milky Way and in fact are nearer to us than the average faint stars of the Milky Way, and secondly that the nebulae, especially those which are unresolved into individual stars, lie outside the Milky Way, which is itself essentially composed of stars. Nebulae are Milky Ways in their own right, which of course corresponds closely with the view put forward by Kant. In spite of this, or perhaps because of it, astronomers were loath to accept this theory of the universe; argument persisted to the end of the century, and the last phase was not reached until the early 1920's.

The Dumb-bell nebula in Vulpecula. Nebulae of this type, evidently composite bodies, are regarded by Hoyle as the biggest problem in present-day astronomy. They may give a clue to the formation of nebulae

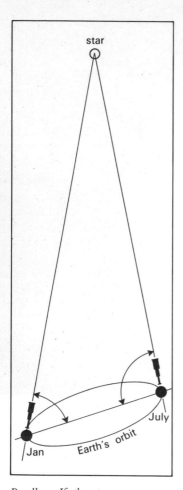

star

July

Jan

Earth's orbit

Parallax. If the star were a very great distance away, the two angles would add up to 180°. For nearer stars, however, the angle sum will be slightly less and this enables the distance to be worked out in terms of the diameter of the earth's orbit

The opposition to the idea of 'island universes' scattered through space, of which the Milky Way is the one in which we find ourselves, was based primarily on the view that the nebulae were very small compared with the Milky Way. Of course, the small size is explained now by the very much greater distance of the nebulae than was thought by astronomers at the time. Another argument, on the basis of an economy of hypotheses, was that, according to Harlow Shapley's work (1919) on the dimensions of the Milky Way, there was plenty of room in it to accommodate even the fairly large spiral nebulae. Although these arguments persisted through the 1920's, a more satisfactory means of determining the distances of the nearer spiral nebulae was developed which settled the question.

It is necessary to say something here about how astronomers determine distances. It is particularly important to see how such determinations are carried out in cosmology, because the very concept of distance is something that is only made precise when a rule is given for determining it. Distances of near objects are determined in very different ways from those of far objects, and a little care is needed in deciding whether the concept of distance should really be counted as the same in each case. For distances that are not too great, a method know as parallax is applicable. If a star is observed from two points of known separation, the angle specifying the star's position will not be quite the same from each point. From the length of the base-line and the difference of the angles, the distance of the star can be calculated. All we need for this purpose is a suitable base-line; the greater the distances to be determined, the longer the base-line we need, for otherwise we shall have to measure two angles so close together as to be almost indistinguishable.

The earth's surface provides a suitable base-line for determining the distance of the sun, and by a series of measurements throughout the year it is possible to construct the earth's orbit very accurately. Once this is done, the most suitable base-line, and the largest we have, for observing the other stars is the major axis of the earth's orbit. To use this, one needs to make observations of the position of a star from the same observatory at midwinter and midsummer. The theoretical basis of the method is just the same; only the scale is altered.

The distances of nebulae proved too great to be determined by the parallax method, but just how much too great remained undetermined. Quite another method has to be used to determine their distances, and the new method determines a concept which certainly overlaps distance as determined by parallax, but cannot be assumed to be identical with it.

A number of stars had been found within the scope of parallax measurements which all exhibited a curious and cyclic variation of intensity. Moreover, the average brightness of each one of these stars, the Cepheid variables, in any given period, varied inversely as the square of its distance away. Such a variation is exactly what would be expected due to the distance, if all the stars of one period had the same absolute intensity.

Now by 1922 some variable stars had been identified in a spiral nebula and a year later the first Cepheid-type variable was found in the nebula in Andromeda which was catalogued as M31. The Cepheids in M31 had perfectly typical variations of luminosity, and so it was reasonable to assume that they had the same absolute intensity as the Cepheids near to us. On this assumption, the distance of this nebula could be determined from the observed intensity of the Cepheid variables.

Parallax measurements are limited to distances of about 160 light-years, but inside that range are numerous Cepheid variables. These can be identified out to about 5 million light-years, giving the distance of a few galaxies, in particular M31 in Andromeda, and of the brightest stars in them. By comparing brightest stars in galaxies, such as M81 (above), the distance scale can be stretched to 25 million light-years

light years   160   5m   25m

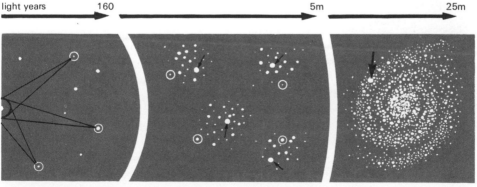

Opposite, top to bottom: the continuous spectrum of white light; the beautiful auroral light, caused by excited atoms and molecules in the upper atmosphere; the characteristic yellow lines emitted by heated sodium; the spectrum of sunlight, crossed by black absorption bands. The sodium pair is easy to spot and indicates its presence in the sun's corona

To resolve light into its constituents by interference the waves may be passed through two slits so that, depending on their phase, one wave can amplify or annul another

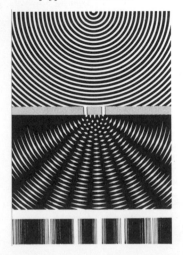

A factor that restricts the widespread use of this method is that Cepheids and other variable stars tend to be fairly faint, so that it is not possible to resolve the more distant nebulae sufficiently to observe Cepheids in them. The assumption made by Edwin Hubble in 1923–25 in judging the other nebular distances was that the brightest stars in any nebula are of much the same absolute intensity. With this assumption one has only to resolve the brightest star in a nebula and compare it with the observed brightness of the brightest stars in, for example, M31; the fact that the star in the other nebula looks fainter is then attributed to its greater distance.

The first tremendous consequence of this determination of the distance of nebulae was a complete confirmation of the 'island universe' hypothesis. The distances of the nebulae turned out to be many thousands of times greater than any single stars of our local galaxy. It was concluded that they must be systems of stars like our own, spread out in space and of the same general size as the Milky Way itself. But a more bizarre consequence was soon to follow. To explain this it is necessary to go into more detail about the measurements astronomers make of the spectra of the stars.

Information conveyed in light

When white light is passed through a prism, it is broken up into its constituent colours because different colours of light are bent through different angles in passing through the glass. This is the same effect as is responsible for the rainbow, and the beautiful stretched-out range of pure colours, from red through orange and yellow to green, blue, and violet is called the spectrum. In fact a prism does not provide the most convenient means of analyzing light, and astronomers prefer other related optical effects, for example, interference, which is the cause of the vivid colours seen when a thin film of petrol is spread over water. But the result is the same.

Now when atoms of a particular chemical element are excited by heat, for instance, light of only specific

colours is emitted. Correspondingly, when the spectrum of the light is examined, there is no longer the continuous band of colour as in the case of white light, but instead a few sharp lines occur at the points of appropriate colour, and the rest is in darkness. Similarly when white light is passed through a particular gas, specific colours are absorbed by the gas. These are the same colours as those which it would emit on being heated, so the spectrum of the white light is crossed by dark lines, or absorption lines, where these colours would otherwise occur. And so, by studying the spectrum of a star, the astronomer is able to determine what elements are emitting light in the star and also, by the dark lines, what absorbing materials lie between the star and himself.

When the spectra of the more distant nebulae were observed, they exhibited the same general pattern of those of nearer stars, but with this curious difference: there was a shift of all the spectral lines towards the red end of the spectrum. For reasons that will soon be clear, this is usually attributed to the motion of the nebula away from the earth. On the basis of observations of a few such nebulae, Hubble was able to say that the amount of red-shift was proportional to the (Cepheid-determined) distance. There are other nebulae which have not been resolved by telescope; for these, therefore, no determination of distance is available. In this case it is assumed to be consistent with Hubble's law, so that here, for the most distant matter, yet another concept of distance comes in – distance defined by red-shift. With this insight, the way was open to map the heavens on an unprecedented scale, and the result was a picture of the universe which corresponds roughly with that described in the first chapter of this book. We shall see later, however, that Hubble's red-shift law had other equally important implications. In the past few years, objects have been discovered with red-shifts so large that, on the Hubble interpretation, they must lie near the very edge of the observable universe. These objects have now become of utmost cosmological importance.

The Hubble relationship between red-shift (expressed as speed of recession) and distance for a number of nebulae in different constellations. They lie on a straight line in the graph because the speed is proportional to distance

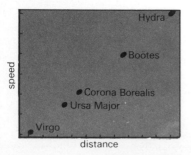

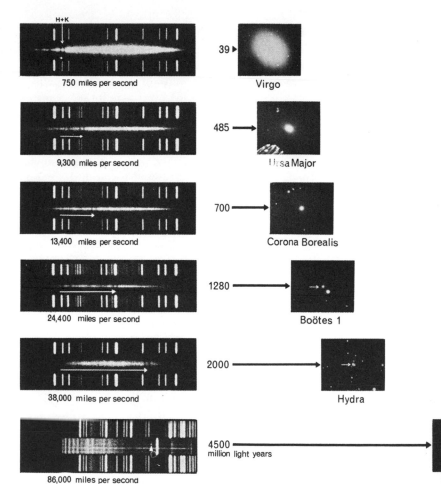

H+K

750 miles per second

39 ▶ Virgo

9,300 miles per second

485 → Ursa Major

13,400 miles per second

700 → Corona Borealis

24,400 miles per second

1280 → Boötes 1

38,000 miles per second

2000 → Hydra

86,000 miles per second

4500 million light years → Boötes 2

We are now almost ready to take the next great cosmological step forward. We have an almost modern picture of the distribution of matter in the heavens, but so far no theories to explain how it has behaved over time. This is one of the great cosmological problems, for any theory of the universe must explain both how the universe behaved in the past and how it is likely to behave in the future. But cosmology is an intellectual jigsaw puzzle and there is still one piece missing. Before we can move on we must assimilate some great advances made in physics in the nineteenth century. These concern the nature of light.

The spectra of the same nebulae, showing increase of red-shift with distance. The upper and lower spectra are laboratory ones, for comparison. The white arrows show the extent to which the two calcium lines are shifted. The spectrum of 3C295 is shifted so far that these lines disappear off to the right, and other lines have to be used

59

An understanding of light

Towards the end of the nineteenth century a satisfactory theory of light as an electromagnetic radiation had been found. The discoverer of this theory is perhaps more responsible than any other man for the immense wealth of new discoveries in physics in the present century. He was James Clerk Maxwell, born in Edinburgh in 1831. In 1856 he became Professor of Natural Philosophy in Aberdeen, and went in 1860 to a similar post at King's College, London. It was in this period that he roughed out his electromagnetic theory. He was transferred to Cambridge in 1871 as the first Cavendish Professor of Physics and in 1873 his great book on electricity and magnetism was published.

Maxwell's theory describes so exactly the observed phenomena of optics, that we can state with virtually no possibility of error that light is simply one of the forms of solution of Maxwell's equations. The consequence of this fact is very important and to understand it we must look more closely at Maxwell's work. The theory of the attraction and repulsion between electric charges has been known for a long time, and by the nineteenth century it was well established that two charges, for example of the same magnitude e and distance r apart, repel each other with a force proportional to e^2/r^2. We can only say 'proportional to', since no unit of charge has so far been defined. But we could so define the unit of electric charge that the force was equal to e^2/r^2, and this would then be the electrostatic system of measurement. Maxwell completed the unification, begun by Michael Faraday, between this theory and that of magnetism.

Historically, magnetism began by discussing the attraction between portions of magnetized iron, but by the nineteenth century it had begun to be apparent that the magnetization of iron, though easy enough in practice, was a very complex matter theoretically, and not a suitable basis for a theory. But once electric currents had been discovered, a connection between electricity and magnetism soon appeared. First, it was found that an electric current flowing near a compass

James Clerk Maxwell, the creator of the electromagnetic theory of light.

needle would deflect the needle from its normal north-pointing position. Second, it was found that two electric currents flowing near one another exerted a magnetic force on each other.

This did provide a satisfactory means of defining magnetism – in terms of the force produced between the two currents. For two parallel currents, the attractive force was related to the currents and their distance apart, just as was the electrostatic repulsion between two charges. If the currents were of value J, and they were distance r apart, the magnetic force between them would be proportional to J^2/r^2. The electrostatic force was, of course, proportional to e^2/r^2. There was an obvious and appealing relationship between the magnetic and the electrostatic systems of measurement.

This should not surprise us because an electric current is no more than an electrostatic charge moving uniformly through space or along a wire. If the charge moves at velocity V, then it produces a current of eV. The magnetic force of attraction between two such currents will then be e^2V^2/r^2. In other words it appears that in order to change from the electrostatic system to the magnetic system we must multiply by the square of a velocity. Although we can go further, it becomes too complicated to explain here. The normal unit of electrical force can in fact be converted to the unit of magnetic force by multiplying by the square of the velocity of light.

Again, it is no surprise that the speed of light occurs in Maxwell's complete theory – in which light is described in terms of varying electrical and magnetic fields. It is also reassuring to find that we now have two ways of measuring the speed of light, and that both give the same answer. The first is to compare the size of the magnetic unit of force in the laboratory with the electric unit. As we have seen, these two units have a relationship to one another determined by the square of the speed of light. The second method is to measure the speed of light directly by timing the path of a light ray over a measured distance.

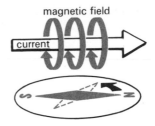

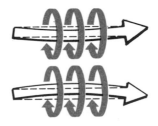

Above: the magnetic field lines round a current are concentric circles. As a result a compass is swung from its normal position. Below: two parallel currents exert an attractive force on each other as a result of their magnetic fields

All this appears fine until we go back to Newton. His laws were found to apply in all inertial frames of reference – that is, they would work equally well in a 'stationary' laboratory and on a smoothly running passenger train. Naturally, it should be expected that Maxwell's laws also apply under these conditions. But here there is a problem. To calculate the magnetic force between two currents, a knowledge of the speed of light is necessary. This might be easy enough in the laboratory but what about the situation on the train? Maxwell's laws say nothing about whether the speed of light is to be taken as the speed in the direction that the train is moving, at right angles to it or in the opposite direction.

Now a man walking at 3 miles an hour along a train moving at 60 miles an hour actually covers the ground at 63 miles an hour. Naturally, we would expect the speed of light from the back of the train to the front of it to be the speed of light plus the speed of the train. But if we work out the magnetic force between two currents on this sort of assumption, we arrive at an answer that is bigger than the one obtained in the 'stationary' laboratory. Not only does this defy common sense, it defies observation. When the experiment is made on the train, it is of course found that the measured magnetic force between two currents is exactly the same as the one measured in the laboratory. What has gone wrong?

Philosophically, it is too appalling to consider the possibility that Newton's laws will work only under one set of conditions and Maxwell's under another. The only way out of the dilemma is to recognize that there is something curious about the speed of light. There would be no problem if the speed of light were regarded as an absolute constant – in other words, if it were unaffected by the speed of the light source or by the speed of the receiver. If that were so, Maxwell's laws would work in all the inertial frames in which Newton's do.

While this idea seemed contrary to common sense, it was soon to be tested. It was Maxwell himself who hit on the idea of measuring the velocity of light in the direction of the earth's rotation round the sun, and comparing it with the velocity of light at right angles to this; and it was the American physicist, A. A. Michelson, who first tried the experiment in 1881 and repeated it six years later with the help of E. W. Morley.

There were enormous experimental difficulties. By terrestrial standards, the earth moves round the sun at enormous speed – some 66,600 miles an hour. But this is equivalent to only $18\frac{1}{2}$ miles a second, and the velocity of light is 186,000 miles per second. If any difference was to be detected in the speed of light moving in the direction of the earth's rotation and at right angles to it, it would be very small indeed. In fact, the difference in time taken by the light to travel the paths set up in the experiment was expected to be only about 1 in 100,000,000. Nevertheless, the experimenters managed to set up equipment which was sufficiently accurate. They found that if there was any difference at all in the velocity of light travelling in the two directions, it was less than one-fortieth of the difference that could be explained by the additional speed of the earth in one direction and not in the other.

This experiment showed that the laws determining the speed of light were something quite different from those determining the speed of, say, a cricket ball. The speed of light was a constant, regardless of whether the source of light or the receiver of the light was moving. While this absolved Maxwell's laws from the difficulty discussed above, it opened up a whole range of new problems.

It took the genius of Albert Einstein to realize that the answer lay in the apparently unconnected problem of how to measure time. This, as we shall now see, led him to formulate his theories of relativity which incorporated a new theory of gravitation. Our knowledge of the universe and our theories of cosmology were soon to take a startling new turn.

RELATIVITY

The next great step forward was due to one of those astonishing combinations of theoretical advance and experimental observation which occur from time to time in the history of science. The theoretical advance, which we shall discuss in this chapter, was concerned with the union of two great branches of physics – mechanics (including the theory of gravitation) and optics. This union of the sciences arises over the question of what the conditions are under which we can say that two events happening a long distance apart are happening at the same time. At first it seems like making unnecessary difficulties to ask questions like this. But 'time' is really only a convenient name that we use in allotting numbers to events (i.e. the 'times' at which they happen). We may do this for events that are very near to us by means of a wristwatch, a waterclock or one of the most accurate atomic clocks; but no matter how accurate the timepiece, we are still only arranging the events near to us in a certain order. The order of events a great distance from us needs to be constructed in some other way.

Einstein was the person who realized that the concept of simultaneity of distant events was by no means as simple as it had seemed and that it had to be defined. In doing so, he elaborated in 1905 the so-called Special Theory of Relativity. This theory does indeed unite optics and mechanics, but excludes the theory of

Opposite: Einstein at home in Princeton in 1954 at the age of 75. Princeton had been his home since 1933 and most of the last 20 years of his life were spent in a vain attempt to find a theory that would unify the diverse parts of modern physics – gravitation, electromagnetism and quantum theory

gravitation. The corresponding theory of gravitation is called the General Theory of Relativity. In order to understand this, however, we must first understand the special theory.

We return, then, to Einstein's realization that the notion of simultaneity of distant events is one requiring definition. In fact, only when one could verify it by some operational procedure could one say that two events were simultaneous. This operational procedure necessarily involves transmitting information from one event to another or from both of them to a common observer. If one starts to elaborate the procedure it becomes clear that the sort of signals which one employs are going to be important. One could employ various signals such as sound or light. We know that the effects of these are not the same and that using two signals at once can lead to apparently paradoxical situations – as when one watches a piledriver from a distance and sees the hammer hit the pile before one hears the sound accompanying it. Of course, such paradoxes are easily explained away by saying that the speed of sound is different from that of light. But at the same time they show that we ought to fix upon some particular signalling system.

Light provides the fastest and most convenient means of signalling that we know but it also occupies an important position in special relativity due to the fact, revealed by the results of optical experiments, that the travelling speed of light is by no means such a simple concept as the speed of a cricket ball. Einstein realized that the two problems, that is, the strangeness of the speed of light and the need for definition of distant simultaneity, could be resolved in one experiment by using light as the signalling method.

Let us imagine first a very simplified situation in which an observer P who is at rest in an inertial frame is making observations on another observer Q who is moving uniformly away from him in another inertial frame (a space-ship perhaps). Let us suppose that they synchronize their watches as they coincide at the same spot. Now suppose P sends light (or radio) signals

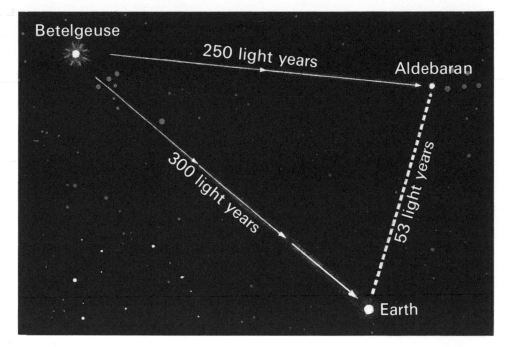

The diagram shows Betelgeuse, Aldebaran, and Earth with distances labeled: 250 light years (Betelgeuse to Aldebaran), 53 light years (Aldebaran to Earth), and 300 light years (Betelgeuse to Earth).

towards Q and notes the times at which he sends them. They are reflected back from Q, and P also notes the times of return. The act of reflecting back the signal is taken to be an event at Q, so by means of a succession of such signals P can produce a succession of events at Q. P can then allot times to these and other events at Q if he has some rule that relates the emission and reception times with the time that he is to assign to the reflection event.

This rule cannot be determined experimentally, for that would mean that there was some criterion by which to judge simultaneity of events, whereas the whole procedure is required for just this purpose. On the other hand the rule is not completely arbitrary, and obviously the time assigned to the reflection-event should be somewhere between those of emission and reception. Moreover if both observers assign the time zero to the moment at which they synchronized their watches, then if P sends a new signal when the time of emission of the first has doubled, he will get it back

The simultaneity of events not too far apart, say on the earth's surface, is clear. In space the situation is different – light from an explosion on Betelgeuse (for example) would not arrive on the earth and on Aldebaran simultaneously, because of the enormously different distances, and each would allocate a different time to the event

at twice the time at which the old one was received, for Q will then be twice as far away. It seems natural for the rule to specify that the time for the new reflection-event is also twice the time which it assigned to the old one. By general arguments of this kind it is possible to decide on a unique possible rule for assigning time to distant events, and it is a very simple one – the time to be assigned to the reflection event is to be the average of the times of transmission and reflection of the signal. We shall call this 'Einstein's rule'.

So far everything has been described from the point of view of P. But since P and Q are in uniform relative motion, they must give completely equivalent descriptions of all mechanical phenomena, since they are both at rest in inertial frames. Suppose that Q is carrying with him the same kind of clock as that used by P. To be specific consider a signal sent by P one second after P and Q were coincident, and let us suppose that the reflection of this signal has returned to P. According to Einstein's rule the time that P assigns to the reflection event is the average of the time of emission (i.e. one second) and the time of return. Does this mean that Q will record the time of arrival of the signal from P as this average value? We must evidently be careful here. We said above that Einstein's rule is not an experimental fact but a convention; whereas what Q measures is an experimental fact. So the rule give us no reason to suppose that Q will measure the signal as arriving at the time assigned by P.

On the contrary, a little careful consideration shows that such a supposition flies in the face of common sense. For suppose that we look at matters from Q's point of view; and imagine that, as he cannot conveniently transport high-powered transmitting apparatus in his space-ship, he contents himself with reflecting P's signal. Let us pretend, for the sake of argument, that the times measured by Q for the reflection events do agree with those assigned by P on the basis of Einstein's rule. Since P and Q are meant to be completely equivalent, in all respects, the same must be true the other way round; that is, the times measured by P for events near him must be the same as those assigned by Q on the basis of Einstein's rule. We shall find that it is impossible for both these states of affairs to hold, and so the only way to preserve the symmetry is to suppose that neither holds.

The time of return to P of the signal which was sent at one second depends on how far away Q has gone, and so on the speed of separation of P and Q. To fix ideas, let us suppose P and Q are separating at such a speed that the signal returns to P at a time 4 seconds after P and Q separated. Einstein's rule is then applied by P to give a time assigned to the reflection event of the average of 1 and 4, that is $2\frac{1}{2}$ seconds. Since we are pretending that Q measures the same time as P assigns, this means that a signal is reflected at Q and starts off on its return journey to P at a time measured by Q as $2\frac{1}{2}$ seconds.

Next suppose that P is a rather cooperative observer, and that as soon as he receives back this first signal from Q he transmits a second one. Since this happens at 4 seconds after the coincidence of P and Q, that is, four times as long as the time when he sent his first signal. P and Q will be four times as far apart and the corresponding time of reflection at Q will be four times the former time of $2\frac{1}{2}$ seconds, that is, 10 seconds. Consider, however, that the first reflected signal from Q to P and the consequential one transmitted from P to Q, can be looked on as a single signal transmitted from Q reflected at P and received again at Q. Accordingly Q may, with equal justification, use Einstein's rule and assign to the P reflection-event the average of $2\frac{1}{2}$ and 10 seconds, that is, $6\frac{1}{4}$ seconds. But we already know that this event was measured by P as taking place at four seconds! So we see that the assigned time exceeds the measured time by a considerable amount (by something over 50 per cent of the measured time in fact). Since the assumption 'P's assigned times are equal to Q's measured ones' leads to the conclusion 'Q's assigned times are greater than P's measured ones' we have a plain contradiction to the fact that P and Q are equal in all respects.

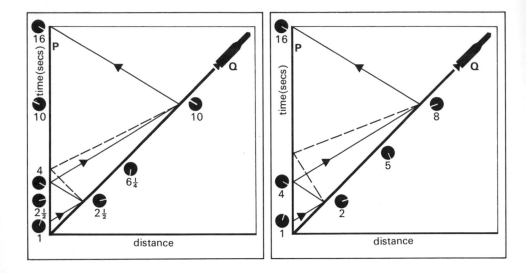

Above: light signals passing between P and Q, moving away from each other in inertial frames of reference. Left: the times for reflection events marked along Q's path are those calculated by P according to Einstein's rule. Accepting these times, and applying the same rule, Q calculates (dotted line) a different time for P's transmission of the second signal. Right: applying the rule to his own recorded times, Q still fails to calculate emission time. The solid line links times recorded by both observers

Einstein's contribution, of course, was to specify the ways in which P and Q differed. He proved that if two observers are moving with respect to one another at speeds large compared with that of light, they will measure times differently, and the size of the difference will depend on their relative speeds. As we shall see in a moment, in the example we have been discussing the relative speed is $\frac{3}{5}$ that of the speed of light. And Einstein showed that the assignment of time would then be 25 per cent longer at distant places than the times measured locally. In other words P would assign the first reflection at Q to $2\frac{1}{2}$ seconds but Q would actually measure it at 2 seconds; P would assign the second reflection at Q at 10 seconds but Q would time it at 8 seconds. By Einstein's rule, Q would then assign the return of the signal to P at the average of 2 and 8, that is 5 seconds. But P has timed that event at 4 seconds, and the 25 per cent difference appears again.

It should now be clear that there is a unique way of relating assigned times for distant events with locally measured times which agrees with common sense. But this is only half the problem. We must now turn to the other half – that is, the question of the strange properties of the velocity of light. In the numerical example Q was receding from P. Since P assigns a time of $2\frac{1}{2}$ seconds to the first reflection-event, he likewise assigns a distance away of Q at that time of $(2\frac{1}{2} - 1)c = (1\frac{1}{2})c$, where c is the speed of light; for this is the distance light can travel in the prescribed time-of-flight of the signal. Now Q has reached this distance in a time of $2\frac{1}{2}$ seconds (according to P), so that, in all, P will assign a speed of

$$\frac{(1\frac{1}{2})c}{2\frac{1}{2}} = \frac{3c}{5}$$

to Q (and because everything has been made symmetrical Q will assign the same recession speed to P).

Suppose now that Q makes observations of another space-ship R which is receding from Q in the same straight line as P but in the opposite direction, and with the same speed $\frac{3}{5}c$. Faced with this situation Newton would have had no hesitation in stating that R was separating from P with speed $(\frac{3}{5} + \frac{3}{5})c = \frac{6}{5}c$. Here, however, he would have been quite contradicted by common-sense, for if R were receding with a speed $\frac{6}{5}c$, signals from P would never catch up, so that R would be unobservable from P. This cannot be so, since Q can communicate with R and signals need not originate at Q but could be the ones coming originally from P. Moreover it is quite clear that we could pursue our analysis in just the same way as before and so find the correct speed of R relative to P; without any calculation, however, it is obvious that it will be less than that of light. Thus, in defining time for distant events in a common sense way, we have also incorporated the peculiar limiting character of the speed of light and this is just what is needed in connection with the difficulties over optical experiments.

But if these difficulties are triumphantly overcome, it must be admitted that this is only at the expense of some affronts to our natural prejudices. And the most discussed of such affronts is that of the ageing of space-travellers. Suppose that we are dealing with three observers P, Q and R but that now R is travelling with speed $\frac{3}{5}c$ in the direction towards P, having started at

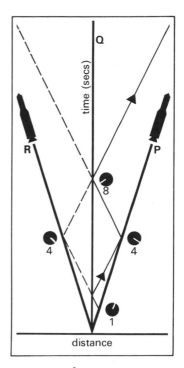

Travelling at $\frac{3}{5}$ c, P and R separate from Q in opposite directions. A signal sent from P at 1 second will return from R at 16 seconds. P calculates R's speed as $7\frac{1}{2}$ c (i.e. half 16–1) divided by $8\frac{1}{2}$ (i.e. reflection time assigned by P), which is $\frac{15}{17}$ c, not $\frac{6}{5}$ c, as might be expected

a great distance, whilst Q is still travelling away from P as before. Let us begin with P and Q who have as before synchronized their clocks at zero at the instant when Q was just passing P. To give the example more point we can use the same numbers as before but suppose now that they represent days instead of seconds. After one day P transmits the first signal; let us suppose that Q receives this just at the same instant that he is coincident with R's returning space-ship. He is therefore able to tell R that (by his clock) he is 2 days journey out from P, and so R can confidentially expect to be passing P 4 days after zero hour. However, P assigns the time $2\frac{1}{2}$ days to this midpoint encounter, and so he can, equally confidently, expect to receive R 5 days after he saw Q depart.

This state of affairs has come to be known as the 'clock paradox' though it is obvious that there is no paradox here. But the situation described is really intended as an idealization of the one in which a space-ship blasts off at zero hour, turns round at a very great distance after travelling for two days at a speed $\frac{3}{5}c$, and returns to earth. If the direct effect of the enormous

The clock paradox. If R sets his clock by Q, in space, he will record the time of his arrival at P as 4 days. P, however, assigns a time of $2\frac{1}{2}$ days to the encounter between Q and R, and therefore times R's arrival as 5 days. The observer moving from rest in an inertial frame and returning to rest at the same point in it ages more slowly than the observer who stays at rest in it all the time

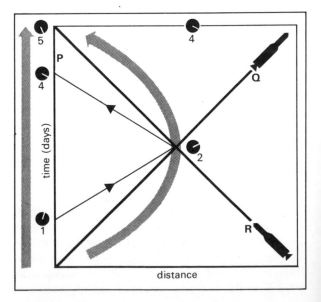

accelerations needed from the rocket motors can be neglected, the same calculation shows that the space-travellers will gain a day's life relative to the stay-at-home. Several things need to be said; in the first place, it certainly is possible to arrange accelerations in such a way that there is no direct effect on the time-reckoning. Secondly, there is no inconsistency or paradox in the theory predicting that the traveller ages less than the other observer. For there is no symmetry between them; the rest observer remains at rest in one inertial frame the whole time, but the traveller does not. At best he can only claim to spend (almost) all of each half-journey in one inertial frame.

Thirdly, there is the question of what is observed, as distinct from what the theory predicts. As far as biological mechanisms are concerned, there is so far no experimental evidence that the ageing process must be considered governed by the same laws of time as physical process. Since ageing is really a matter of cell destruction and regeneration, it would be very surprising indeed if it were governed by a different time-variable from other physical processes. But if anyone wishes to stick to such a view, in order to prevent differential ageing of space-travellers, there is at present nothing to refute it.

The matter is quite different with purely physical processes, however. Here there is ample experimental evidence of the so-called time dilatation, and the most straight-forward evidence comes from cosmic rays. The soft component of the showers observed on the earth's surface are mainly the particles known as μ-mesons (mu-mesons), which are charged particles whose masses are about 200 times that of the electron. These particles are well known in the laboratory. They are unstable (though their lives are longer than those of most other elementary particles) and their lifetimes, measured in the laboratory, are about two millionths of a second. Now these μ-mesons are mostly produced in the upper atmosphere, approximately six miles up. Even if they were to travel with the speed of light, ordinary Newtonian mechanics would allow them to

travel only some 600 yards in their lifetimes. In fact they do travel with a speed very near to that of light; and accordingly their ageing is slowed down. We saw above that for a speed $\frac{3}{5}c$ ageing is slowed down by 20 per cent; similarly for speeds nearer to c the reduction is correspondingly greater. For the μ-mesons a reduction of 90 per cent or more is to be expected and this explains completely the observation of so many at sea-level.

The special theory of relativity, which is essentially what we have been describing, was published by Einstein in 1905. The mathematical details, in one particular specialized context, were in fact in print in the previous year, by Lorentz, and in 1905 a paper also appeared by Poincaré. It is fair to say, however, that these two scientists were looking at only one specialized, detailed point of view. It was Einstein who realized the general physical significance of the transformations.

Mechanics and relativity

Our next step is to combine all this with mechanics and when we have reconciled these ideas with mechanics in general, to turn our attention to by far the most important piece of mechanics which we have discussed up to now, that of gravitation. The first question we must consider in mechanics is, what modification needs to be made to the definition of the acceleration in Newton's theory. We have one sure way of starting here; we know that, at least for velocities very small compared with that of light, Newtonian mechanics is accurate. Accordingly, however we may generalize our mechanics, we must be sure that the small velocity approximation is Newtonian.

On the other hand, there must certainly be some changes when the velocity gets bigger. For the acceleration can be related, either to the extra velocity produced by an acting force or to the change in velocity that results. These two concepts, which are completely identical in Newtonian mechanics, must be different here. When a force acts on a body an

Opposite, left: recorded on photographic emulsion carried high into the atmosphere by a balloon, a collision occurs between two nuclei, producing a shower of mu-mesons travelling at nearly the speed of light. Right: mu-mesons decay in about two millionths of a second, in which time they can travel about 600 yards. But at such high speeds time slows down for the mu-mesons, allowing them to travel ten times as far, reaching the surface of the earth in considerable numbers

additional velocity is compounded with the original speed of the body. But we know that compounding this extra velocity may make virtually no difference if the speed is already very near to c. Now Newton introduced the idea of a numerical constant called mass to measure the extent to which a body resisted the action of forces. Since changes in velocity are harder to produce when the velocity is near to that of light, we can expect that this will show up in an increase in the apparent mass when the speed is large.

It is really a matter of convenience whether we use the word acceleration to measure the extra velocity compounded, or to measure the rate of change in speed. The most convenient, however, turns out to be neither of these, since they both depend in a complicated way on the speed the body has already. This is awkward because original speed is measured relative to the observer; and so the speed (and therefore the measure of the acceleration) can be altered by simply changing the inertial frame of the observer.

In this problem there is at each moment one observer whose description of the motion of a particle is a particularly valuable one at that time. This is the observer who happens to be moving with the same speed as the particle at that time, so that for him the particle is at rest. His description is particularly valuable because we know that the lower the speed, the more accurately Newtonian mechanics apply. Accordingly the best possible use of Newtonian mechanics is in the inertial frame in which the particle is at rest – the rest-frame. We can then define the acceleration to mean the Newtonian acceleration measured in the rest-frame. Of course it is not necessary always to construct the rest-frame at each instant in order to work out problems in mechanics. By means of a small amount of algebraic manipulation we can easily find formulae to determine the motion of a particle under given forces in any inertial frame. But notice here the limitations; we must be using an inertial frame and in that frame the forces must be given. Subject to such limitations it now remains to find the expression to take the place of the momentum mv of a particle of mass m moving with speed v in Newtonian mechanics.

The best way to find cut what should take the place of the momentum is to consider the impact of two different massive particles, comparing the accounts of this impact given by two observers in uniform relative motion. When we do this, we find that the momentum has to be re-defined as

$$\frac{mv}{\sqrt{1 - v^2/c^2}}$$

Such an expression agrees completely with the one already in Newtonian mechanics, mv, when v/c is very small, but at higher speeds it begins to give a higher value of the momentum. If one measures mass by impact experiments, which is one of the ways in which masses of elementary particles are found, this increase in momentum will give rise to the expected increase in the measured mass. The measured mass will be

$$\frac{m}{\sqrt{1 - v^2/c^2}}$$

and so will no longer be a constant, as Newton assumed.

Here again our prejudices are faced with conflict though, because of the very limited character of our everyday experience, there is no conflict with common sense. We often think of mass rather in Newton's terms as the 'quantity of matter in a body', but in deciding on the numerical value to be given to it, the only criterion is the experimental method of determination. Not only is the formula for mass increase

Champion's 1932 experiments on the collision of electrons with stationary electrons. Left. a relatively low energy collision. In this case the angle between the outgoing tracks is very near to 90°. At higher energies (right) the particles tend to come off at a narrower angle. This is caused by the increase of mass with velocity

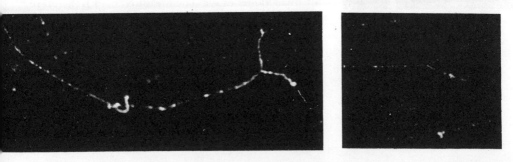

Beneath the earth embankment (above), nearly a mile in circumference, the giant synchrotron accelerator at Serpukhov, USSR, accelerates protons to speeds near that of light. Below: on the right is part of the vacuum tube in which the protons are accelerated to their final high energy

completely justified by measurements of the tracks of fast-moving particles in cloud-chamber photographs; the mechanics of particles moving in the circular rings of big particle-accelerators is also worked out on this basis. One could certainly say that even the smallest defect of special-relativity mechanics in this region would be extremely expensive in terms of millions of dollars!

It is to be noted that, as v approaches c, the measured mass becomes larger and larger, so that larger and larger forces are needed to accelerate particles already moving at high speeds. This is another aspect of c as a limiting speed; not only is it the same speed for all observers and that of the fastest signal we know at present, it is also a speed beyond which it is impossible to accelerate a particle, because of the way in which its mass increases with velocity.

What is the origin of this increase of mass with velocity? We can get a clue if we notice that

$$\frac{1}{\sqrt{1 - v^2/c^2}} \simeq 1 + \frac{1}{2}\frac{v^2}{c^2}$$

for small values of v/c. The measured mass is therefore approximately equal to the constant mass (corresponding to zero velocity) plus a term proportional to the kinetic energy E, or $\frac{1}{2}mv^2$,

$$m' = m + \frac{E}{c^2}$$

By rewriting this approximation in the form

$$m'c^2 = mc^2 + E$$

Einstein was led to interpret the term $m'c^2$ as the 'total

energy', made up partly of kinetic energy E, and partly of 'rest-energy' mc^2 which the particle had even at rest. It follows from this interpretation that if only a small amount of the 'rest-mass' could be destroyed in some way the resultant release of energy would be very large indeed because of the factor c^2 (in cgs units about 10^{21} cm^2 sec^{-2}); and we have seen the enormous consequences of this, for evil and good, in the atomic bomb, and the nuclear power station. These modifications of mechanics are also of vital importance in the discussion of gravitation, to which we must now return.

Gravitation

Here it is appropriate to remember the limitations under which the generalization of mechanics has been made. We needed to be using an inertial frame, and to know the forces. These conditions are fulfilled, to a very good degree of approximation, for charged particles moving in an electromagnetic field in the laboratory. But in the case of the gravitational field there are no adequate experiments for distinguishing the inertial forces in one frame of reference from those in another; to be more precise, it is not possible to distinguish gravitational forces uniquely from other forces produced by transformation to an inertial frame. Everything gets mixed up. This problem held Einstein up for ten years after his formulation of the special theory of relativity in 1905.

We usually talk, as we explained before, not about the gravitational force between bodies – because this is a rather inconvenient thing to deal with – but with

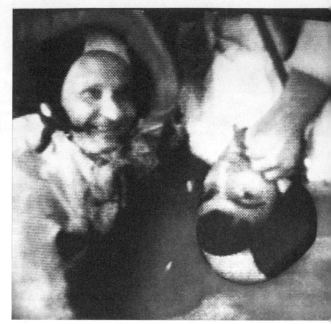

the 'gravitational field' surrounding a body. That is to say, if we are, for example, near the surface of the earth, we find that any particle, released from rest, will fall with a certain acceleration. This acceleration, as we saw before, does not depend upon how massive the particle is. This is quite a different situation from the electro-magnetic case, in which the acceleration of a charged particle in an electric field depends upon its mass and its charge. A heavy body accelerates just as quickly in the gravitational field as a light one. Of course, in everyday circumstances this is obscured to some extent by extraneous circumstances like air resistance. However, we have nowadays far more weighty evidence for the universal quality of the gravitational field which we have been talking about than bodies falling near the surface of the earth. Astronauts have actually experienced the phenomenon of weightlessness when they were in a satellite freely falling in the gravitational field. What they find is that not merely some of the bodies in the satellite are weightless and others almost so or that some are weightless on one occasion and others on another occasion: they find that when the satellite is freely falling everything is weightless. The effect of the gravitational field has been removed completely by looking at it from the point of view of a frame of reference which is freely falling in the field. This can be done because of the universal character of the gravitational field, the fact that the acceleration that it produces in a body does not depend upon how massive the body is.

This fact had been recognized as important by Galileo and indeed, in the later development of mechanics it had been used to simplify the mathematics for the description of the motion of the planets in the same way as we did in the last chapter, but it had not been made to enter fundamentally into the theory. The emphasis rather, in Newtonian gravitation, was on the inverse square quality of the field which we also discussed there.

For the sake of simplicity, let us look a little more closely at this universal quality of gravitation, putting on one side the necessity already mentioned in this chapter for modifying our ideas of distant simultaneity and the transformations between inertial frames. Let us look at the whole situation from a purely Newtonian point of view and, to fix the ideas, study one particular problem (studied also by Galileo) – that of the motion of a projectile. By throwing a stone into the air you will at once observe that it follows a curved path. As a matter of fact the curve is one well known to geometers, under the name 'parabola' and the properties of this curve were used by Newton and his successors to work out the theory of the projectile. Nowadays the properties of this curve are not so well-known. It is interesting to observe that the universality of the gravitational field and the ability to transform it away allow us to solve problems about projectiles equally well by performing this transformation. That is to say, taking our cue from the astronaut we look at matters with respect to a frame of reference which is falling freely.

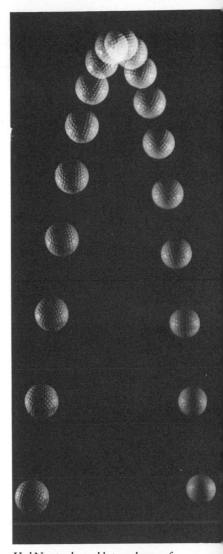

Had Newton been able to make use of stroboscopic photography, his study of the motion of a projectile (in this case a golf ball) would more easily have revealed the parabolic trajectory. The deceleration while leaving the ground is equal to the acceleration while approaching it

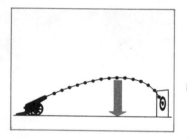

Above: the trajectory of a shell from the point of view of an observer fixed relative to the cannon. Below: the situation from the point of view of a freely falling observer

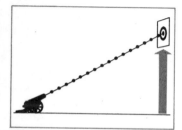

Let us set ourselves one particular problem, that of firing the projectile with a given muzzle velocity so as to hit a target which, for the sake of simplicity, we suppose to be on the same level as ourselves. The corresponding figure in the freely falling reference frame is very much simpler. Since the system is freely falling there is no gravitational field in it and the projectile in this reference frame moves in a straight line with uniform speed. The only complication is that the target, instead of being at rest, is now moving upwards with a uniform acceleration but this makes it simple to calculate the elevation on the cannon, knowing the time it takes the projectile to reach the target.

There is another way of looking at this, which is helpful because it stresses the way in which the effects which are usually described as 'physical' have been carried over into the mathematical description of the background. In the freely-falling reference frame the possible paths of the projectile are straight lines; and these are, of course, the basic units (apart from points) from which Euclidean geometry is constructed. Corresponding to any geometrical properties of straight lines there must, from what we have said, be other properties (in general much more complicated) of parabolas – for the latter properties will simply arise by transforming the former ones. So instead of performing the transformation we could start with this new kind of geometry, suitably formulated, in which the part of a straight line in ordinary Euclidean geometry is played by a parabola. We could then describe the parabola as the 'natural' path in the geometry and so conclude that the projectile moved along it under no forces – a rather daring generalization of Newton's first law.

Matters are not quite as simple as this because time enters into it: the fact must be included that the projectile describes a straight line at a constant speed. This can be done by a slight extension of the idea of geometry, by starting, not with the point as in the usual geometry books, but with the event – a point associated with an instant of time. If we want to continue to draw pictures of the geometry on a piece of paper, which has only two dimensions, we need some technical devices. These are akin to perspective, which is the device that enables us to make a two-dimensional arrangement of points and lines look like a three-dimensional solid (when we have been trained to look at it in this way). Of course it is rather complicated, so it is easier to abstain from drawing pictures at all except in the much simpler case when all the events are in a straight line i.e. in one dimension, which we can take as horizontal on the paper. Then we can use the vertical line on the paper to mark off, on a suitable scale, the time of the events, so that the points on the paper are a picture of a set of events.

Now the set of events corresponding to successive positions of a particle moving with constant speed in a

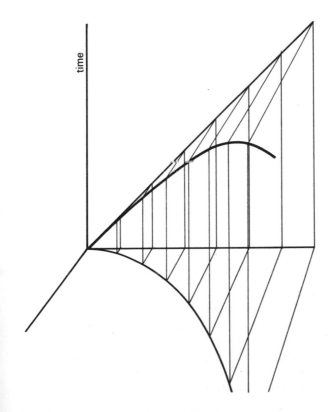

Time is measured in an upright direction in the diagram, and the flat plane is that in which the motion takes place. For simplicity, the figure shows the path in the flat plane traversed by a shell projected horizontally, that is, along the line running across the page. If gravity did not act (freely-falling observer), the shell would continue along this line. If gravity acts in the flat plane in the direction towards the reader, the path is bent into a parabola. The successive positions of the shell along the line, or along the parabola, are at successive moments of time, so in the three-dimensional figure the occupation of each position by the shell is an 'event'. The time of the event is represented by the height of the corresponding point above the position (making a three-dimensional curve). The positions along the curve then lie on an inclined straight line of events

straight line will be represented in this picture simply by a straight line, the slope of this line depending on the speed of the particle. When we transform to the parabola picture, the new way of looking at things means we have to employ three dimensions (one for time) and in this picture the parabola is stretched out into a rather complicated three-dimensional curve. But this complication does not affect the theoretical possibility of 'geometrizing' gravitation in this way.

It is clear from this example that we can abolish the gravitational field by a transformation, although of course this is not a transformation between inertial frames, like those discussed in the last chapter. It is an accelerated transformation. We have here the possibility, as a famous American physicist puts it, of 'gravitation without gravitation'. If one can reformulate all the rest of one's physics in a way that does not change when one makes transformations between frames of reference, then any problems concerning physical phenomena in a gravitational field can be solved by first transforming to the frame of reference in which there is no gravitational field, then solving the physical problem, and then transforming back to the original frame of reference.

Is gravitation just a hoax, then? No, not entirely. There are two difficulties here. In the first place, one has to be able to formulate the original problem in a manner which is not changed when one goes over to the new frame of reference which is accelerating relative to the old one. This can be a tall order, though it is fair to say that the difficulties involved in it are usually only of a mathematical character; that is to say, it depends only on the ingenuity of the physicist performing the formulation. Much more serious is that the discussion of the projectile so far applies only in a uniform gravitational field. In fact gravitational fields are only very approximately uniform. The field near the surface of the earth is uniform, but as soon as we bring the orbits of the planets into consideration, the inverse square law field applies. There is no accelerated transformation which cancels the effects of the inverse

square law field, because if one performs the appropriate transformation at one point or in the immediate neighbourhood of one point, then, as soon as one moves away a little, the field is changed and therefore must be accounted for again.

The original Newtonian gravitational field is, in fact, somewhat of a hoax, because it can be transformed away at a point, but what really concerns us is the way that this field differs between two points and that is certainly not a hoax. The idea of the gravitational field is, as it were, shifted one step along. What we had before, we can cancel out, but we cannot cancel out the way it changes from point to point. All of these ideas could be used to re-work Newtonian mechanics in a way which would be interesting and satisfactory. What concerns us here is something rather different. Einstein used this approach to gravitation to connect it with the special theory of relativity which he had

Transforming away the gravitational field is a purely theoretical exercise, but it has a bearing on practical problems. Pilots ejected from supersonic aircraft can be subjected to multiple gravity, like this US Air Force 'guinea pig' undergoing a stress of 22g in a rocket sled

found in 1905. Putting aside an unfinished version of this in 1911, he finally reached the complete form of his theory, the general theory of relativity, in 1915. All that this theory does is essentially the following: it starts with the kind of description which we have given for inertial frames, in the first part of this chapter, that is to say, one in which the time is involved in a rather complicated way with the spatial co-ordinates, unlike the Newtonian situation, and it poses the problem of relating this to making accelerated transformations so as to transform away the gravitational field at one point.

What transformations will now be permitted between the space and time co-ordinates? The answer is in some ways extremely simple and from other points of view terribly complicated, because it is a fact that the new space and time co-ordinates can depend on the old space and time co-ordinates in almost any way whatever (putting on one side certain conditions of continuity, i.e. requirements that nearby points have nearly equal co-ordinates, which are naturally assumed). And since these are the transformations which one is to perform, the rest of one's physical principles have got to be written in a way which is not changed by these transformations. This is known as the Principle of General Covariance of the theory.

Before going on we had better be clear just what is involved in this. The main purpose in talking of a gravitational field in Newtonian theory, instead of considering always gravitational forces between pairs of particles, is that it is a great aid in the formulation and solution of complicated problems in the theory. Consider for example the problem of the figure of the earth. Simplifying this drastically, we could ask the question 'What shape is assumed by a rotating mass consisting of an ideal incompressible fluid?' Of course the real earth is not composed of such a fluid, but this simplified problem will suffice to show the role of the gravitational field. Since the fluid is rotating it is pretty clear that it will be roughly spherical, but flattened at the poles (as indeed the earth is). But by how much?

The accurate calculation must balance the effect of the rotation, causing the bulging round the equator, against the gravitational attraction of one part of the fluid on the other. This attraction cannot be exactly determined, however, until we know the shape of the mass. The mathematical theory affords a way out of this impasse, and the technique depends on formulating a gravitational field theory satisfying certain equations and then solving these equations simultaneously with those determining the equilibrium distribution of the matter. In all this theory it proves possible, in Newtonian gravitation, to specify the field everywhere by giving the value of a number everywhere, the so-called gravitational potential. (The strength of the actual field at a point is then determined by how fast the potential changes when one moves from that point to another nearby, and its direction is determined by the direction in which one should move to make this rate of change greatest.)

Now at first sight general relativity seems a very different theory because the gravitational field is there to be described by the background geometry. But it turns out not to be so different after all. As the geometry changes from point to point there are certain quantities called the ten metric coefficients which vary with it and describe, by their values, the modification of the geometry. In the particularly simple case of a particle moving slowly (compared with the speed of light) in a weak gravitational field (that is, any of the fields we ever encounter in Newtonian gravitation) one of these metric coefficients differs only by a constant from a multiple of the corresponding Newtonian potential for the problem. In a more complex case (fast motions or stronger fields) all ten of the metric coefficients will be needed.

Having done all that one has to write down the appropriate field equations for the metric coefficients which will determine, not the original Newtonian gravitational field, for that can be transformed away, but the way in which the corresponding thing to the original field changes from point to point. (It is useful to notice here that, since we are going to be interested in the way that the quantity like the Newtonian field changes from point to point, it is not the changes in the metric coefficients that will be important for us, but the changes in those changes – a cause of some complexity in the theory.)

Now the field equations have to be determined in such a way that the field corresponds to the inverse square law field of Newton rather than a law of force varying with distance in some other way. When the mathematical details of this are carried out one can then search for solutions of the theory, that is, values of the metric coefficients, and compare them with the experimental situation. For many years, very few solutions were known and the most striking confirmation was in the case of the orbit of Mercury. As we said in the last chapter, the orbit of Mercury is not an ellipse, as Kepler's law suggests, but a slowly rotating ellipse, the rotation being about 5000 seconds of arc per century. All but 50 seconds of arc per century are explained by the gravitational effect of the other planets and it was to explain the remaining 50 seconds that the planet Vulcan was invented in the last century. Unfortunately, however, no such planet was ever discovered in the predicted orbit. When one uses the field equations of general relativity to describe the motion of a single planet round a fixed sun one finds not an ellipse, as in Newton's theory which explains Kepler's law, but a rotating ellipse, and, in the case of Mercury and the sun, by about 50 seconds of arc per century.

We must, of course, be clear just what is being explained here. One finds 50 seconds of arc in this idealized case and one guesses that the complicated system of the real world with the other planets is adequately described by adding together the relativistic rotation of the orbit of Mercury and the Newtonian calculation of the effect of the other planets on the orbit. Let us be fair to general relativity, however; it is often stated that the explanation of the rotation of the orbit of Mercury is one of its few experimental confirmations. It should be clearly understood that even if general relativity had only predicted the same gravitational results as Newton's theory, which it certainly does in its first approximation, it would still have been a vastly superior gravitational theory for it reaches these results with far fewer arbitrary assumptions than Newton's theory. The field equations proposed by Einstein, although they give rise to a field which changes like the inverse square law field, have much more secure foundations than Newton's; they are indeed almost the only field equations it is possible to write down under the limitations of the theory.

One may wonder what made Einstein reformulate gravitational theory in a manner which, although appealing to the theorist and providing much material for the mathematician, seems to be somewhat perverse compared with the simple Newtonian formulation. His motive was quite specific. He was very much concerned with the problem of the existence of preferred inertial frames in the universe. It was noted in the last chapter that one can determine a local coordinate system, in which Newton's laws have a simple form, by means of a pendulum, that is, by measuring the rotation of the plane of swing of a spherical pendulum, and when one does this one finds that the distant matter in the universe does not have any transverse motion relative to this frame of reference. This must mean that the distant matter has some effect on the choice of the local inertial frames of reference. This fact, the so-called Mach's principle, puzzled Einstein for many years. He saw that, for example, the Newtonian theory of gravitation was unsatisfactory in that, when it described the orbits of planets as ellipses, it did so relative to an inertial frame, but it left out of account the fact that the experimental frame was one fixed relative to the distant stars. Only half of the phenomena had been explained, and what one needed was a gravitational theory in which the solution of the problem of a planet moving round the sun was

Opposite: Mach's principle. When Foucault demonstrated the rotation of the earth he was really showing its rotation relative to a Newtonian inertial frame. Such a frame, however, is not rotating relative to the distant stars, and somehow these must be determining the local inertial reference frames

possible only when the effect of all the most distant matter in the universe was also taken into account. This would be a theory, according to Einstein, which 'incorporated Mach's principle'. General relativity was intended to be such a theory.

Argument has raged ever since about the extent to which general relativity does incorporate Mach's principle. Some people believe that it does, at least in part. Most take the view that it does not incorporate Mach's principle at all. However that may be, Einstein at once set about finding solutions of his field equations which would apply to the universe as a whole and not merely to the solar system. Of course, when one wishes to find solutions to field equations applying to the whole universe one does not try to deal with the whole universe in its genuine complexity, consisting of separate nebulae and clusters of nebulae or separate stars. Rather, one looks for a description of a smeared-out distribution of matter which will have the right density on the average and which is easy to deal with by the kind of mathematics used in general relativity.

Einstein was not able to find a solution of his original field equations which corresponded to the actual universe. He ran into exactly the same problem the Newtonian cosmologists had run into: having an infinite universe and therefore infinite gravitational forces at every point. Because of the nature of his equations, however, he was able to see the only simple modification which was possible and which would enable him to have a model of the universe which, as he believed, incorporated Mach's principle. He had to introduce an additional term into the equations, but because of the very rigid nature of general relativity this additional term was essentially the only one he could introduce. When he introduced it, the resultant static universe was the so-called Einstein universe.

The Einstein universe
Whatever its advantages and defects as a model of the actual world, the Einstein universe was a milestone in the path away from the infinite universes and infinite

gravitational fields that plagued cosmology for so long. It can best be described as follows: imagine that we are concerned only with two dimensions, so that our classical cosmological model is a plane, infinite in all directions. The step which Einstein took corresponds to considering instead the surface of a sphere. There is still no boundary – a straight line (i.e. shortest-distance-between-two-points curve, or great circle, in the geographer's parlance) can continue indefinitely (though it will come back and join on to itself eventually) – yet the surface has a finite total area, unlike the plane. So in the three-dimensional case, the curvature introduced into the geometry means that the Einstein universe has finite total volume, though no boundary.

Unfortunately the addition of the extra term in the equations, the so-called cosmological term, was not sufficient to incorporate Mach's principle. Whatever the status of Mach's principle in the original theory, it was essentially left unchanged by the extra term. Shortly afterwards, Willem de Sitter found a solution to the field equations in which there was no matter at all and yet the space was not the field-free space of Newtonian physics when matter was absent, but a more complicated kind of geometry. This was a great blow to Einstein's hopes of incorporating Mach's principle and, as a result, in later years he gave up adding the cosmological term to the equation, since it did not achieve the desired result. The problem of Mercury had been solved by looking for a solution of the equations found by Einstein which corresponded to a particle representing the sun and nothing else at all. For the purposes of the cosmological problem, Einstein wanted to find solutions in which there were no particular points at which masses were concentrated, because it was the smeared-out version of the universe that concerned him. Instead of looking for spherical symmetry, he wanted a more complicated situation in which the view from one point in any direction was much the same as the view from any other point. These requirements on the model are usually known as the cosmological principle. There are, in fact, only

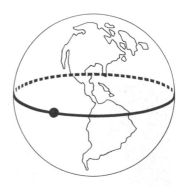

Einstein's universe possessed in three dimensions the same 'closed' property that the surface of the earth has in two dimensions. It is possible, on the earth, to start off straight along a circumference and, without turning, return to your starting-point. Similarly for Einstein's universe

three solutions of the equations in conformity with this principle, which satisfy the requirement that the solution does not depend upon the time. One of these is the field-free case of no gravitation at all, in which the empty universe is simply that described by special relativity. One of them is Einstein's solution. The other is de Sitter's. An additional embarrassment for the Einstein solution was the fact that, although it represented a static configuration of matter, this configuration was shown to be unstable.

The de Sitter universe

The de Sitter universe was a difficult one to conceive of because of the absence of matter, but if we imagine that the actual solution given by de Sitter is probably a good approximation to a real universe in which the matter is very thinly distributed, then we can think about the solution and find that, in such a universe, the matter a long way away from us (although it has no transverse motion, because of the frame of reference which we have chosen) is receding from us along the line of sight with a speed proportional to its distance away. Indeed, if we introduce a small particle into the universe, and make the assumption that its introduction will not affect the universe as we have found it, we discover that the particle will accelerate away from us acquiring a speed proportional to its distance.

If the de Sitter universe is to provide any model for the actual universe, we shall have to think how the universe could be expanding in this way. If all the matter is moving apart with speeds proportional to the relative distances, it must, a long time ago, have been very tightly compressed. What could have been the nature of this original universe? Questions like these begin to raise what is really the fundamental problem of cosmology, and one which was not clearly formulated until the 1940's by Hermann Bondi, Fred Hoyle and Thomas Gold. If the original universe were very much more compressed than the present one, how do we know what laws of physics to apply to it?

Above: Willem de Sitter. Below: de Sitter's universe has no matter in it, so is hard to depict. But if one imagines a few particles (on the central triangle) to act as 'markers', these separate from each other at speeds proportional to their distances apart

The laws of physics which we have at the moment are those devised on the surface of the earth in conditions which, to an astronomer, are extremely atypical. If we want to talk about the universe a long time ago and to suppose that it was then very compressed, we have absolutely no idea which of our physical laws, if any, we can retain unchanged in this new situation and which are approximations corresponding to our present situation. Once we realize this dilemma we see that a theory of cosmology is much more difficult than we had imagined. We can, if we like, make the assumption that all the physical laws are unchanged, even in extremely different physical situations. But we shall have to regard the results of this assumption with considerable scepticism.

The idea of all the matter in the universe receding from the earth with a speed proportional to its distance sounds at first very shocking indeed. Has man again been elevated to the position in the centre of the universe that he had for Aristotle? Fortunately, the answer is no. The diagram overleaf shows that if all the matter is receding from us with speed proportional to distance, and we survey things from another point instead, we should again find all the matter receding from that point with a speed proportional to the distance. If the experiments had shown some other law of recession, say, with speed proportional to the square of the distance this would have been a very serious matter. The way in which the de Sitter universe gives rise to this recession may perhaps be better illustrated by an analogy. Imagine again that man was one of a race of little two-dimensional creatures living on the surface of a sphere. This solution would, in fact, be the one found by Einstein, the surface of the sphere having no boundary and yet being finite in total area (in Einstein's actual solution, in total volume). The de Sitter universe corresponds, in the analogy, to little beings living not on the surface of a fixed sphere, but on that of a balloon which is being blown up. Each of the beings is receding from the others with speeds proportional to their distances measured along the

The evolutionary, or 'big-bang' theory of the expanding universe assumes that a long time ago all the present existing mass of the universe was concentrated in a much smaller region. The expansion has resulted in the present distribution

surface, but none of them is in a privileged position. As a matter of fact, a velocity proportional to the distance is the only one which is consistent with the universe's looking the same from every point.

No one had ever tried to construct a cosmological model of that kind before. A static solution had always been sought. It should be made clear that the de Sitter solution does not correspond to the sort of gravitational field or the sort of space which depends upon the time because of the expansion. Everything remains the same at any later time as it is now, but if any matter were introduced into the solution without disturbing it then that matter would rapidly acquire a speed proportional to its distance from us. It is in this sense that the de Sitter solution represents an expanding model. Naturally, the discovery of such a model led cosmologists to look at the solutions of the equations which depended upon the time taken for the expansion. It was conjectured that the universe might perhaps have started off in a static Einstein state which was disturbed, and then moved through a succession

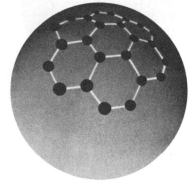

The universe expands in such a way (velocity proportional to distance) that every point has an equal right to be considered the centre of the expansion. An analogy may be drawn with a number of points marked on the surface of an inflating balloon

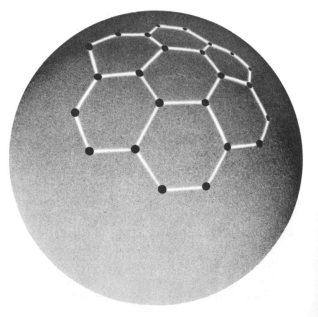

of models corresponding to its expansion, finally finishing up after a very long time in the de Sitter situation. But speculation of this kind needs, above all, experimental facts to verify it, and these were not long in coming.

Experimental verification

The relationship between theory and experiment is, however, most complicated. We can simplify matters a little by looking at a few of the dates at which events occurred. After discovery of the de Sitter universe, it was an obvious step for the theoretical astronomers to look for some generalization. It had already been shown that there were only three static models (empty space, the Einstein universe and the de Sitter universe) which satisfied the theory of general relativity. Accordingly, if there were to be any description of the universe which was more accurate than those described above, it would have to be one which was not static. In 1922 A. Friedmann discussed the possibility of a space which was curved in much the same way that Einstein's universe was curved (like the surface of a sphere rather than a plane). This was what general relativity required for the analysis of mechanical phenomena, but now this curvature was assumed to be dependent upon the time. In other words, there was a general resemblance to our description of the de Sitter solution, only now the difference was that the varying curvature allowed matter to be present in the model.

It was necessary for Friedmann to make some assumption about the way in which the curvature varied with the time in different places, and he made the obvious one that the averaged-out picture of the universe as a whole was represented by a space which at any instant had the same curvature at every point. In fact, such an assumption will follow from certain simpler and more plausible assumptions, in particular that, looking from any point, space seems the same in every direction. However, Friedmann's two papers on this subject, in which a great deal of the later theory of what we now call the expanding universe was worked out, were totally ignored by almost all his contemporaries. Whether it was that the mathematics seemed to offer too great a difficulty or whether the papers seemed at the time unlikely to lead to anything of astronomical interest is not now clear, but in any event Friedmann's work was known very little until nearly ten years after its publication, and then only as a result of the interest which the works of the Abbé G. Lemaître and H. P. Robertson produced.

Lemaître in 1927, in complete ignorance of Friedmann's results, developed a remarkably similar theory, and Robertson in the next year, again independently, produced the same mathematics. The sort of models derived in this way, the so-called Friedmann-Lemaître models, consist of a universe which is expanding in the sense that the average matter in any region is moving away from the average matter in any other region. The amount of curvature of the space at any point decreases with time. In this way the observed expansion of the nebulae, which was to be found in the following year, 1929, by Edwin Hubble, was satisfactorily accounted for up to a point. It is not altogether clear to what extent Hubble was influenced by this theoretical work, but it seems that he was working fairly independently of it, and that the observational activities on the other side of the Atlantic were very little influenced by the theoretical developments in Europe. In fact, as in the case of Friedmann's theory, the much fuller development of Lemaître and Robertson was largely ignored by astronomers and by the scientific world in general. All these models of an expanding universe were little known until a paper by Sir Arthur Eddington appeared in 1930. Eddington had been working with a pupil of his at the time, George McVittie, on the problem of the instability of the Einstein solution. They came across Lemaître's paper and realized at once how an expanding universe gave them the solution to their problems. For it showed what happened to the unstable Einstein universe once it had been disturbed and moved from its initial 'over-full' state.

The recession of galaxies, accurately measured in 1929 by Edwin Hubble (shown opposite with the 48-inch Schmidt telescope) made the present expansion of the universe an observable certainty

The later mathematical developments of the theory of the expanding universe proved a happy hunting ground for mathematicians, but not altogether with useful physical results. Once the expansion was assumed, there were certain quantities at the disposal of the mathematicians. It turned out that, by a suitable choice of these various quantities, one could have models which expanded either from a state of extreme compression or from the Einstein state. Moreover, they could continue to expand indefinitely, or contract after a certain time back to a state of compression, after which they might begin to expand again (the oscillating universe).

In an intermediate state some regions of such a model would have begun to contract. There are interesting reasons why some such models may be ruled out. These reasons are connected with the so-called arrow of time. Most physical laws are reversible ones; that is to say, if we film a certain dynamical experiment, and then run the film in reverse order, there is nothing to tell us that we are not watching a film

Various cosmological models are possible according to general relativity. Some expand and then contract (1), some do the opposite (5), still others expand continually, either from a singular state of immense compression (2, 3) or from an initial Einstein state (4)

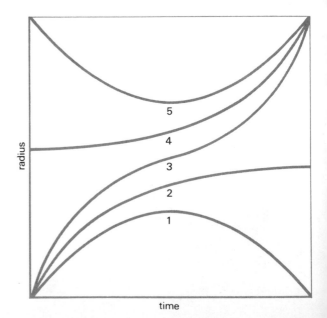

run correctly but of a different experiment (the experiment, in fact, in which the initial and final conditions of the original experiment have been interchanged).

Sometimes this situation is rather different, and there is irreversibility. We know that a film which shows a puff of smoke growing smaller and disappearing into someone's mouth, followed by a cigarette, is being shown in reverse. Such an irreversible phenomenon is an example of the arrow of time. Death is another, although, being biological, it is more complex. Again, we always receive television programmes a short time after they are transmitted, although the usual theory of their transmission, which takes no account of the whole universe, would equally allow the reverse possibility of always receiving them a short time before they were transmitted. Most people feel there is a connection between the arrow of time, as evidenced by these local effects, and the expansion of the universe. If this is so, and a contraction of the universe leads to a reversal of the arrow of time, then we would expect to discover in contracting models some very strange phenomena in distant places. In these regions where the contraction has already begun, as it were, disorder decreases instead of increasing. It is difficult to imagine how we would be able to integrate the description of such regions into that of the rest of the universe which was still expanding. None the less, such an oscillating model has been proposed and studied in some detail by R. H. Dicke in more recent years. We shall return to this later. In any case the final decision must, of course, be made from observation.

The purpose of observation in these circumstances becomes to try to decide what values to give to the constants which occur in the model so as to make it agree as closely as possible with the actual universe. But here, of course, we are up against an extremely difficult practical problem, for we only observe the actual universe at one moment. It has not been possible to correlate past observations over any considerable amount of time, and therefore any model which has a period of expansion is bound to fit fairly well. A great deal of work was done on establishing the validity of the fundamental assumption on which the model was built, that is, that the universe is roughly the same looking in all directions from any point. In fact, it is impossible to decide from such observations on any one of these models in favour of another, at least for a very long time to come.

If we formulate different models of this kind and then expect some experimental choice between them, we are obviously forgetting the extreme difficulty involved in making measurements in cosmology. Astronomers have, over the years, managed to make measurements of quite remarkable accuracy, but here we are considering extremely large regions of space – so large that light takes a considerable time to reach us. This involves us in long time-lags since the light left the original nebula, and thus any real discrimination between theories with only marginally different predictions becomes quite impossible. There is also a difficulty of quite a different order. We have said in the first chapter that cosmology is the theory of how the universe comes to be as it is. One essential feature of it is, therefore, that it is the discussion of a unique system. It is not, like other branches of science, the discussion of a whole range of systems, some of one kind, some of another, varying in some respects but similar in others. We have to have, for a theory of cosmology, a unique answer. The relativistic cosmologies gave a whole range of answers depending on how one chose the various constants in the expansion theory. From this point of view, rather paradoxically, the old static cosmology of general relativity was more satisfactory, for here it was possible to show that there were only three possible models; two were completely empty and therefore unsatisfactory as models of the universe, and this left the only one containing matter, the Einstein universe. But of course, although philosophically satisfactory, one could not hold to the model of the Einstein universe; for one thing, the density of matter in it was far too great, and for another it did not predict any expansion, and expansion was observed. This situation is unfortunately one which occurs all too often in cosmology; the theories which are philosophically satisfying are difficult to match with the observations, whereas those which fit with the observations prove to be philosophically rather naïve.

A serious attempt was made to overcome this problem towards the end of 1970 by William McCrea. He set out to formulate explicitly a philosophical position for evolutionary models instead of taking a particular view for granted. Briefly, McCrea's view was that the theory of the universe ought to take account of the increasing difficulty of gathering information at great distances, somewhat as quantum mechanics does for the difficulty of information about very small entities. He made certain proposals for doing this but it is probably too early yet to pronounce on their correctness, or on whether they are sufficient. What was important was McCrea's acknowledgment that an explicitly and clearly formulated philosophical standpoint is essential in cosmology and not a kind of luxury as in most other sciences.

Olbers' paradox

In order to see how the subject has developed since 1930 it is instructive to look back at an earlier semi-philosophical argument, which again has been very influential in recent years. This is now generally known by the name of Olbers' paradox, although it was known very much earlier than 1826 when Heinrich Olbers investigated it. In fact, very much the same argument had been produced by le Chevalier P.L. de Chéseaux in 1744 and the paradox had been hinted at as early as 1720 by Edmund Halley. In an extremely simple form the paradox is as follows. We observe that the night sky is dark, except where there are stars. The stars, of course, give out a great deal of light but this spreads out in a uniform manner in all directions from each star, so that the greater the distance from the star, the greater the area of the sphere over which the light is spread. The amount of light which is available in any given small area on such a sphere therefore decreases with the distance from the star. Now assume that, on an average, all the stars (as in Olbers' original argument although we would now replace this by nebulae) have roughly the same magnitude, and suppose that they are distributed in a uniform manner so long as we take the average over large enough regions of the universe. Then the number of stars which appear is proportional to the area of the sphere, centred on the earth, on which they lie. We have seen, however, that the amount of light which we receive from each of these stars varies inversely as R^2, where R is the distance from the star. Accordingly all the matter at a distance R from an observer contributes an amount of light locally which is independent of R, since the weakness of the light from the distant stars is exactly compensated by their larger numbers.

If then we add up the contributions from the matter at all distances, this simple calculation gives an infinite brightness in the night sky, in complete contrast with the observed fact that the sky looks quite dark. This theory is clearly a little too simple. If we consider a uniform distribution of nebulae, it is obvious that we shall not see them all because some will be hidden by others between us and them. The total amount of light observed at the earth is therefore somewhat less than that predicted by the previous calculation. More careful calculation shows that the intensity of light which will be received when we go out at night will be roughly equal to that on the surface of the sun, taking the sun as a typical star. This is just as bad as the original paradox. Both de Chéseaux and Olbers decided that the only explanation of the paradox was that there was some interstellar matter of unknown constitution which intercepted the light. However, Bondi rejected this explanation since the matter itself would be heated by the absorption of the light until it reached a temperature at which it would radiate just as much light inwards as it received from the outside. It would then make no difference to the intensity of radiation.

It is noteworthy that neither de Chéseaux nor Olbers questioned the assumption that the stars were more or less the same, and that they were distributed uniformly. We now have much more evidence than they had to show that these assumptions were very reasonable indeed. With hindsight it is easy to see that there is a different way out of the Olbers difficulty. The universe is expanding, and so the light from distant matter is shifted towards the red end of the spectrum by the Doppler shift. But since the frequency of the light signals is decreased, and since in any case the energy of light is proportional to its frequency, the amount of energy which we get is also decreased. When we take the expansion of the universe into account, we find that Olbers' paradox disappears; the

A multiple cluster of galaxies, about 350 million light years away, in the constellation Hercules. The total amount of light emitted by all the stars does not make the night sky infinitely bright, as Olbers thought it should, because the recession of these distant stars reduces the radiant energy we receive from them

night sky should indeed be dark. To look at things the other way round, it would have been possible for Olbers or a later worker to note his assumptions carefully, and to observe that among them there was the assumption that the nebulae were at rest. He would then perhaps have realized that, because of the Doppler shift, this assumption needed to be questioned – which would have led to the prediction of the expansion of the universe merely from the observation of the darkness of the night sky.

It is no discredit to Olbers that he was unable to carry out such an argument, for without the observation of the Hubble law it presented great difficulties of the kind mentioned in our earlier description of the essential difficulty of cosmology. Cosmology, we said, deals with a unique situation: the universe as a whole. It is not its business to compare this universe with other universes; there is only one. It must accordingly be a scientific theory of a unique kind. It is not only difficult theoretically; there seems an irreducible difficulty in the observational side of the theory. Most of the observations are difficult in a perfectly self-evident way. One has to see to the most distant parts of the universe to count the number of nebulae there, determine their luminosity, and so on. We now see that there are other kinds of observation which bring difficulties of their own, although the actual observational side is perfectly straightforward. In Olbers' paradox the observations concerned are made with the naked eye on a dark night; the problem lies in interpreting it. In this particular case it is not too difficult – as long as there is other evidence to suggest the expansion of the universe. But Olbers' paradox is evidently intimately related with Mach's principle, and again the observations which show this are relatively simple. If one takes a reference frame in which Foucault's pendulum does not rotate, one will find that the most distant matter in the universe is at rest in the same frame also. In other words, the laws governing Foucault's pendulum are evidently related in some way to the most distant matter. This observation can be made fairly accurately without any great trouble, but we are still almost completely in the dark about its interpretation.

It would be interesting to make a study of the extent to which this resolution of Olbers' paradox influenced the thought of the Cambridge cosmologists Bondi, Hoyle and Gold, more particularly Bondi, immediately after the Second World War. Certainly the argument plays a large part in Bondi's book on the subject and it probably has had a considerable indirect effect in making the Cambridge cosmologists realize how such philosophical arguments are essential in cosmology. The apparently more down-to-earth physicists, with their expanding universe models, were making other philosophical assumptions of a more questionable kind, without knowing that they were doing so. To some extent the conclusions of the Cambridge cosmologists were anticipated in both 1918 and 1925 by W. D. MacMillan. He was mainly concerned with problems of star formation but, knowing that matter is converted into energy in the interior of the stars, he chose a similar mechanism in reverse to explain away Olbers' paradox. He supposed that, in travelling through space, the radiation somehow disappears, that is to say, is converted into matter. He therefore supposed that some of the radiation from the distant nebulae had been turned into hydrogen atoms, accounting for the observed interstellar matter.

The perfect cosmological principle

This theory is, however, considerably less radical than the one which Bondi and Gold produced in 1948. They put their theory forward from an explicitly philosophical standpoint, or, as they said, they 'made their philosophical assumptions clear rather than concealed'. The first step in their theory is the so-called perfect cosmological principle. We have already noticed above that cosmologists usually make an assumption, often called the cosmological principle. This assumption is that everywhere in the universe is very much the same as everywhere else, which is only

Hermann Bondi (right) and Thomas Gold, scientists who have turned the whole concept of cosmology upside down with the steady-state theory

putting into more precise form the assumption which has always been made in astronomy since Copernicus, that the earth is not the privileged centre of the universe but just a typical planet in orbit round a typical star, the sun. In the perfect cosmological principle, the same assumption that is made about space is also made about time. That is to say, the universe is assumed to be much the same everywhere, and always to have been so. It turns out that this form of the cosmological principle is immensely more powerful than the previous form and accordingly the arguments for both of them need to be looked at rather carefully.

According to Bondi and Gold (we shall come to Hoyle's view in a moment), all the arguments for the narrow form of the principle apply equally to the perfect cosmological principle. They are as follows. Firstly, all physical science makes the assumption that experiments can always be repeated, yielding the same result. In particular, if an experiment is repeated after an interval of six months, when the earth is at a different part of the universe because of its motion round the sun, we should not expect any different result. This underlying axiom requires a certain restriction on the structure of the universe, that is, a cosmological uniformity. This argument becomes much stronger when it is realized that among physical experiments will be included the kind of observations, and their accompanying theoretical discussions, which occur in cosmology. Secondly, as we have said before, in any theory of a changing universe there have to be some assumptions about how local physical laws change when the environment is completely altered. Such assumptions will be completely arbitrary, and progress in this direction is highly uncertain. But in fact, Bondi and Gold argued that such speculation was not required. If the universe is sufficiently uniform both in space and in time, then the situation at a different time and a different place is very much the same as it is here and now, and everything can go on in physics just as we have always known it.

A final argument which they gave for both of these principles is as follows: we have various views about the nature of scientific theories, and philosophers of science are by no means agreed as to which of these views adequately represents the nature of a scientific theory. Probably there is something to be said in favour of all of them. Certainly there is some truth to be found in the view put forward, particularly by Sir Karl Popper, that the nature of a scientific law is that of a hypothesis, and the experiments relevant to it are those which are devised with the intention of refuting the hypothesis. In putting forward this idea of a scientific theory, Popper achieved his object of avoiding the problem of induction. This problem arises if one thinks of scientific theories as collections of facts, together with an assumption that some uniformity among the facts (a 'law of nature') that has persisted for some time will continue to persist. For it is then logically reasonable to ask for evidence in favour of this assumption of uniformity. However, this evidence can only be had by repeating an experiment, and since the principle of uniformity must be assumed before

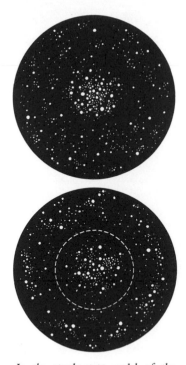

In the steady-state model of the expanding universe it is the observable mass that is constant. As matter recedes beyond the limit of observation its place is taken by newly created matter (within the dotted circle)

assessing the evidence of the experiment, the evidence must be invalid.

From Popper's point of view the best scientific theories are those which give the most opportunity for experimental refutation (so long, of course, as they are not actually refuted by the experiment). We have already seen that the expanding universe theories of general relativity did not offer much opportunity for experimental refutation because they had a number of undetermined constants in them. These constants needed to be determined by experimental means and so, if one did further experiments, one could only determine further constants. Such theories had rather the same structure as the psychoanalytic theories of Freud which always seem to fit the facts, no matter how the facts change. In contrast, the theory put forward by Bondi and Gold on the basis of the perfect cosmological principle was an extremely rigid one. It was very difficult in 1948 to see how it could be altered at all if the experiments were in conflict with it, and after twenty years it still seems almost as difficult.

When the perfect cosmological principle is combined with the observed expansion of the universe, it is clear that new matter must be created in order to keep the density of matter in the universe constant. As the ageing nebulae drift apart it is also necessary for the matter which is formed to condense into new nebulae in the spaces between them. On this basis there will have to be a general distribution of nebulae of all ages. At first sight it seems as if such a theory is going to predict a continual increase of the mass of the universe, which would be in conflict with its own principles. But this is not the case. In any scientific theory of cosmology we must at all times deal only with the observed universe. As the nebulae expand with increasing velocity, the most distant ones which we could observe will be approaching the speed of light. When they are very near to the speed of light, their light is shifted to the red end of the spectrum to a great extent. Less and less energy can be received from them, and when they actually reach the speed of

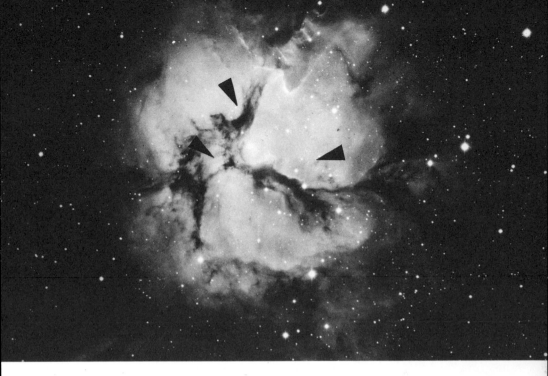

light we will no longer be able to see them at all. When matter is further away than this, it is no longer part of the observable universe. The creation of new matter locally merely provides a constant total mass for the observable universe. From these assumptions it was possible for Bondi and Gold to calculate exactly what the creation rate of new matter needed to be in order to maintain the balance. In fact it amounts to the creation of the mass of a hydrogen atom in a volume occupied by an ordinary sized house once in every hundred million years.

It is to be particularly noted that the Bondi-Gold theory was quite independent of the relativistic cosmologies which have been discussed above. Einstein's field equations for general relativity were not employed. The strong form of the cosmological principle was sufficient to give them everything that they wanted.

The Trifid nebula in Sagittarius, one of a number of regions where new stars are thought to be forming (at the points marked by arrows). Is this formation dependent on newly created matter there? Or does the steady-state model imply a uniform creation process with later local condensation into stars?

Hoyle's theory

At the same time that Bondi and Gold published their theory, Hoyle, who was closely in touch with them in Cambridge, published an alternative theory. At this time Hoyle's views were entirely consistent with those of Bondi and Gold, but in addition he was able to show how these ideas could be reconciled with general relativity. There was for some time considerable discussion about a preference for one or other theory although, looking back after twenty years, the difference between the two theories does not seem big enough to worry about. The really interesting difference is between the predictions of the Bondi, Hoyle and Gold theories on the one hand, and those of relativistic cosmology on the other, despite the glaring philosophical objections of the latter. When considering all the subsequent observational material, one must bear in mind the conflict caused by the two theories. It must be said immediately that it is not yet possible to make a definite decision. At various times in the past twenty years the continuous creation theory, which is usually known as the steady-state theory, has been under extremely heavy pressure from the observationalists. Indeed, so precise and reliable did the evidence become in 1966 about the distribution of radio sources that Hoyle announced that he considered it impossible to maintain the steady-state theory in its original form.

We shall discuss this in more detail in the next chapter, but for the moment it is sufficient to refer to the general idea of radio astronomy mentioned in the first chapter. As it became possible to locate more sources, and to determine their distances, it became apparent that there was a significantly larger number at great distances than would be expected with uniform distribution. Of course we expect more sources to be great distances away, because there is increasingly more room at greater distance from us. But the number increased more than was expected. Since no cosmologist could seriously attempt to explain this by postulating a universe that was non-uniform in space, the only alternative is to have non-uniformity in time, i.e. to reject the perfect cosmological principle. Then the excess of distant radio sources can be attributed to a greater production of them in the earlier stages of the history of the universe. This excess is now seen spatially, at a great distance, because the general expansion has carried the earliest produced matter farthest apart.

On previous occasions the pressure on the theory had turned out to be due to a misunderstanding of the observations or to incorrect observations. On this occasion Hoyle was able to give a more refined theory, which he considered to retain the main spirit of the steady-state theory, and which could be adapted to explain the observations. It would take too long to discuss Hoyle's refinements in detail, which were by no means universally accepted as being truly in keeping with the steady-state hypothesis. Essentially they amount to insisting that, although the universe is in a steady state when looked at on a sufficiently large scale, what we can see at present is a local disturbance. In the course of time it has transpired that the excess of radio sources is less extreme than had been imagined, and, bearing in mind the uncertainty of distance determinations, could hardly be said to be completely inconsistent with the steady-state theory.

Because the theory has managed to survive on previous occasions this does not mean that it will always survive in the future, and, at the time of writing, it is once again under extremely heavy pressure. What the outcome of this will be we cannot tell. But it can be said, as we said in an earlier chapter, that if the steady state theory has to go, then cosmology is in a much worse condition than one imagined before 1948, when the steady-state theory was put forward. For now we realize that we have no knowledge of how to construct a cosmological theory at all, unless it is by making some assumption of uniformity; and the only obvious candidate is the perfect cosmological principle.

The new experimental results will be described in the next chapter but, before we come to that, there is

Fred Hoyle reconciled Bondi's and Gold's ideas with general relativity. In recent arguments about whether observations would still allow the steady-state theory he has been one of its most steadfast defenders

one other interesting piece of evidence in favour of the steady-state theory which may be mentioned. We have already discussed in a general way the arrow of time. In the real world there is a definite element of irreversibility, and one of the ways in which we express this in physics is in terms of order. We define a quantity which represents the amount of order in the system, and this amount of order always decreases.

In our earlier discussion we also took another example of the arrow of time. When we come to discuss the solution of Maxwell's equations for electrodynamics, which we talked about in an earlier chapter, we find that the solution for the field around an aerial of a radio transmitter has such a form that although it describes waves moving out in all directions, we could equally derive from it, by a small modification, a solution describing waves coming in from all directions to the transmitter. It would not matter as far as the equation (that is, the theory) is concerned if the 'film' of the act of transmission were put in the wrong way. Yet in the physical world we know very well that transmitters transmit outwards, and receive nothing. Again, we have seen from observations that our unique universe is actually expanding and not contracting, and this is yet a third illustration of the arrow of time. People have naturally tended to think that there should be some causal connection between these different examples of the arrow of time.

One can see in a general way that the electromagnetic argument ought to be accounted for by the expansion of the universe, for since the universe is expanding it means that energy can be radiated away to the distant parts and there lost. A contracting universe would not allow this indefinite loss of radiation, but would require some to return to the transmitter. That is, the expansion chooses, from the possible solutions of the electromagnetic equations, those particular ones which are the ones observed. When we come to try to work this out in detail in the actual universe, however, it is not quite so straightforward; P. E. Roe has shown that the Friedmann-Lemaître models of the universe do not provide the electro-dynamic arrow of time as one would expect, although the steady-state model does provide it satisfactorily. This affords a strong argument in favour of the steady-state model, either in its present form, or in some modified form.

In such a conflicting situation, the reader may wonder if the right answer lies somewhere quite different. It is possible that both of these predominant schools of thought in cosmology are wrong and that for a correct theory we must make some quite new assumptions. Now because of the nature of the assumptions already made in either the big-bang or the steady-state theory, such new assumptions are virtually certain to involve denying some usually accepted beliefs. Of course this does not rule out such theories; one of the most salutary effects of scientific theories is the way in which they force us to give up cherished beliefs. But it does mean we have very little idea of how to start formulating such theories until we know a good deal about how the existing two are both untenable, if they are. We shall later discuss various suggestions that have been made for such new theories; meanwhile we shall confine our attention to the two main protago-nists.

In the years up till 1960 the experimental observa-tions which were made with the intention of choosing between these rival theories of cosmology failed to produce any definite result. Any piece of evidence in favour of the one theory can be matched with another piece (quite different in kind) for the other. A great deal of work was done but it is fair to say that, to a large extent, more of the same kind of heavenly objects were found as before, and that painstaking observation gradually narrowed down numerical limits on various quantities without revealing anything sensationally new. Since 1960 both the observational and the theo-retical situations have been completely transformed.

It would be fair to describe the situation at the beginning of the 'sixties like this: on one side were the protagonists of the evolutionary big-bang universe model, appealing as always to the common sense and straightforwardness of their approach. They were under heavy siege from their steady-state opponents, who regarded them as naïve in their belief that they were not making metaphysical presuppositions, and who castigated them for providing only a framework within which many models were possible instead of a unique model like the steady-state universe.

However, there has been a great change of opinion, starting in 1961 when Sir Martin Ryle, at Cambridge, announced some results of radio-source counts. We had better be clear about the details of these counts, or else it may surprise us to see how they change. The assumption by Ryle (in the tradition of Olbers, and indeed of Herschel) was that, by and large, radio sources were of much the same intrinsic 'brightness', i.e. magnitude. On the basis of this assumption he was able to give a distance to each one, depending on its faintness, and so to find how many there were at a given distance. The results in 1961 were, on his interpretation, sensational. Of course, on any model, the number of sources at a great distance will increase; for a static homogeneous model it will increase in such a way that the total received radiation from sources of

Above: Professor Sir Martin Ryle. Opposite: partly obscured by a bar of dust, NGC 5128 in Centaurus emits radio noise a thousand times more strongly than a normal galaxy, and remains one of the most enigmatic objects in the universe

any given observed brightness is roughly constant. It is a little complicated to see the effect of the expansion, but it turns out that in the conventional expanding universe model there will be a fairly small falling-off of total received radiation with observed brightness (corresponding to a less rapid increase of numbers with distance, i.e. fewer distant sources). For the steady-state model, this falling-off of total received radiation is considerably greater. But Ryle observed considerably more radiation from distant sources than from near ones – at complete variance with both the conventional expanding universe and the steady-state model. Indeed the steady-state prediction was in error by a factor of 5 or 6. To be sure, there was some scepticism. There had been earlier counts of radio sources which had turned out to be erroneous. However, in Ryle's own words, 'Because we have been wrong in the past does not mean we are necessarily wrong now.'

Such a complete scandal failed at first to budge the

Opposite: used to pinpoint radio sources in space, the three instruments comprising the Cambridge one-mile telescope (top) can be moved relative to each other along a rail track. Signals from the three aerials are combined in receivers (bottom) in the central laboratory. Although much cheaper to construct, the non-steerable radio telescope (below) has to rely on the motion of the earth to scan the skies

theorists significantly, but the radio-counts continued and by 1966 the position had become considerably more challenging. It turned out that Ryle's original figures depended on his having used what was unfortunately a somewhat exceptional radio source as his standard of distance. The 1966 observations (which naturally included many more sources) partly in Cambridge and partly in Australia, can be summarized without going into technical details by saying that the four answers for source counts with distance – 1966 radio observations, static universe, conventional expanding universe and steady-state theory – were still in this order but now with roughly equal gaps between them. Thus, although the conventional big-bang theories were on the wrong side of the static universe compared with observation, the steady-state theory was very much worse off. It was at this time that Hoyle took the view that the steady-state theory was no longer tenable, and set about reconstructing it.

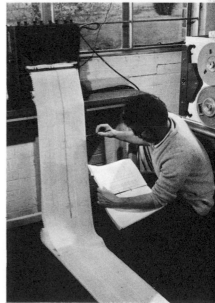

Subsequently the radio-source counts did not diverge more from the steady-state theory than in 1966, though there is no very strong evidence of rapprochement. But interest began to be centred on new evidence of a sensational and exciting kind.

Observations and general relativity

Here, as in so many cases in science, the new observations can only be understood as part of a theory, and we must return now to the theory of general relativity. It had already made predictions before 1960 which caused the observations of that time to be extremely exciting. Let us begin by recalling how Newton formulated his law of gravitation in terms of the force between any two bodies being proportional to their masses and inversely as the square of the distance between them. The important point to notice for our present discussion is that such a force between any two masses is always an attractive one. Moreover, the force always increases as the mass increases. This is in complete contrast to the electromagnetic case. There the force between two charges depends on the product of the charges; but if one considers a random collection of charges then some of them will be positive, and some will be negative; the negative will cancel out the positive and so diminish the resultant force. Besides, saturation effects occur with forces between charges. These are a little more complicated to explain, but they arise because the charges affect their environment to some extent and tend to cancel out their own attractions. No such effects occur with gravitation.

An old-fashioned and apparently rather academic problem about gravitational theory is that of a sphere under its own gravitational attraction but under no other forces. All the particles of such a sphere will attract each other and if there are no other forces acting (by this is meant that the atomic forces between the various particles of the sphere are not to be considered) there is nothing to hold the sphere apart against this attraction. It will gradually get smaller and smaller. The whole process can be worked out

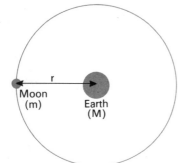

Newton's law of gravitation defines an attractive force (P) between two bodies, proportional to their masses and acting along a line connecting them. The case of the earth and the moon illustrates that $P = GmM/r^2$ (G is the gravitational constant)

quite satisfactorily in Newtonian gravitation and it is a rather curious fact that the time it takes the sphere to contract does not depend upon its radius but only upon its density. A sphere of water of any size in which, if such a thing could be imagined, all the atomic and nuclear forces were suddenly switched off, would contract in about a quarter of an hour, apparently to nothing. Contraction to nothing is extremely hard to imagine and is usually referred to as final catastrophic collapse, though whether the catastrophe referred to is that of the sphere or the theory which is describing it is open to question!

Now imagine that this whole problem and argument is translated into general relativity. Very much the same process is bound to happen at first, because Newtonian gravitation is such a good approximation to general relativity. But we must remember that general relativity is the theory which results when we make Newtonian gravitation consistent with special relativity. Now in special relativity a certain constant, the speed of light c, arose and played an extremely significant part in the theory. In Newtonian gravitation another constant, the constant of gravitation G, is of fundamental importance. Consider now a mass M, which might be our original sphere. It is an elementary calculation to find from the theory that the quantity GM/c^2 is a measurement of length. We are not unaware of situations in scientific theories where critical lengths or times play an important part. Situations in which some such quantities appear are usually those in which the theory begins to have rather peculiar properties or indeed to break down. For example, it was noted earlier (p. 63) that velocities near to c have very strange properties in special relativity. We shall therefore be interested in looking at this critical length which is associated with any particular mass in general relativity.

The first question to ask is: how big is the length? For a body whose mass is about that of the sun the length is of the order of half a mile. In other words, if we could imagine the sun contracting, because all

Solar prominences (above), probably blasts of very hot gas, are visible evidence of the enormous amounts of energy being lost by stars into space. Eventually the fuel for the thermonuclear processes must be exhausted and the stars will cool to absolute zero

the internal forces had suddenly been switched off, as we imagined with the sphere of water, we would expect something peculiar to happen when it reached the radius of about half a mile. This might appear to be a purely academic question. It is surely unlikely that any particular body will reach this critical radius, since there are always other forces acting. To answer this objection we must go back a little and sketch in the physical discussion of the problems of stellar radiation. The stars produce the vast quantity of radiation, which we see coming from them, particularly from the sun, by means of a thermo-nuclear process; in other words, they are all fairly well controlled atomic explosions. But, although we know that the amount of fuel needed for atomic processes is exceedingly small, the thermo-nuclear process must eventually exhaust the material, leaving a star which has been cooled down to the absolute zero of temperature.

Now one can work out what sort of form matter adopts in this case. We have to remember that various kinds of forces are acting on the matter. As well as the gravitational forces, there are chemical and nuclear forces. One can judge the relative importance of these forces by considering small collections of matter.

Matter consists of various kinds of elementary particles. Some of these are of a comparatively heavy kind, such as the proton and other particles of roughly the same mass, whereas some of them, which provide charge, like the electron, are of very small mass. The particles which provide the greater part of the mass are called the baryons or heavy particles. In the present argument we would not go far wrong if for baryon we substituted the word proton. If we take a collection of baryons in order to construct our matter it will probably be charged; of course if they are all protons it will certainly be charged. In order to make neutral matter we must add a certain number of electrons. We can neglect the mass of the electrons in what follows, and so in calculating the gravitational forces the electrons are of no importance.

If such a collection of particles is left to itself it will, according to well-known physical principles, settle down into the state of lowest energy possible, just as a golf ball in a smooth hollow in the ground settles down to a state of rest (i.e. smallest kinetic energy) at the lowest point (i.e. smallest potential energy). This state of lowest energy is called the equilibrium state. For a relatively small number of baryons the equilibrium state has been calculated. In particular, if a collection of 560 baryons is considered, it has a definite unique equilibrium state. It consists of 10 atoms of iron of atomic weight 56, arranged in one particular crystal lattice. The reason that the state is iron rather than one of the other elements, which would be produced by arranging the baryons in a different way, is because of the nuclear forces between them. The particular crystal lattice is specified by the chemical forces. Going upwards, one could take as many as 56×10^{41} baryons but there will be very little difference. The lowest energy state will still be something made up of iron of atomic weight 56. The iron will now be in the form of a sphere of radius about 5 miles.

When one continues to take larger and larger collections of baryons, however, significantly different features begin to emerge. When the mass of baryons amounts to somewhat more than the mass of the sun, the electrons in the central regions become squeezed up to such small volumes that they begin to combine with the protons and make neutrons. If one were to add more to the critical mass, there would be a collapse of the star. The central pressure would rise still more. As a result, more electrons would be crushed into combination with the protons, forming further neutrons; the central regions would contract; this contraction would bring all the parts of the star nearer together; and with the resulting shortening of the distances all the gravitational forces, and so the pressures, would increase. More electrons would be crushed into protons and the collapse would take place faster and faster.

It seems as if there should be a stable end–point to this collapse, when all the protons have been converted into neutrons so that the result is what has now become known as a neutron star. This is not so. The first collapse occurs when the mass is about 1.2 times as great as the mass of the sun. Now when the star has a very high density it turns out that the critical mass at which collapse occurs is only 0.7 times the mass of the sun. So there is no possibility of avoiding further trouble by supposing that the initial collapse ceases with a stable position of very high density. Instead, detailed investigation shows that a second collapse-point occurs with a mass of about 0.7 times the sun's mass. At this point not only are the chemical forces overwhelmed, so that the particles are pushed very close together as before, but the nuclear forces which caused the original model to be made of iron rather than anything else are also overwhelmed and the gravitational force prevails.

Is it not possible to avoid the conclusion that a final catastrophic collapse occurs by supposing that the matter becomes incompressible at a sufficiently high density? No, this is impossible, because at very high densities general relativity is the theory which must be employed, and when one works out the details of a distribution of matter in general relativity it turns out

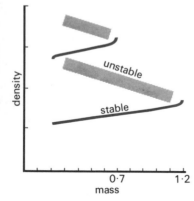

Increasing the amount of matter in a dense star results in an eventual loss of stability at a 'collapse point' 1.2 times the mass of the sun. The rise to a higher density state is shown as a conjectural leap. The process then continues until a second collapse point is reached. The end point for a further rise in density is unknown

that the pressure of the matter contributes an extra term which is added to the mass. When one starts adding small amounts of mass to extremely dense matter under very high pressure, the increase of pressure gives a bonus, and increases the mass still further. In any case, an incompressible fluid in the usual sense is not acceptable in any relativistic theory, because by incompressible we mean that it cannot be compressed at all, and therefore if one pushes on one side of it the other side must respond immediately. Thus an incompressible body transmits signals through itself instantaneously. Such a body is one in which the speed of sound is in excess of that of light, but we have already seen in special relativity that information cannot be transmitted at speeds as high as this.

The second kind of instability therefore seems quite inescapable. One is again presented with the difficult question of what the final state of the system can be, if there is one. Is it possible to imagine that ultimately ordinary matter goes out of existence in some way?

What is being referred to here is not the almost magical pair-annihilation described in quantum mechanics. There an electron and a positron combine in some way so that matter disappears, but the pair is replaced by sufficient energy to make the energy balance correct. In the case of ultimate collapse it seems that the matter really disappears altogether.

The position is, however, a little more complicated in general relativity, because after this final collapse has started, the material must eventually be inside a sphere of smaller radius than the critical length, referred to above, which is associated with its mass. Now it is worth seeing what the general theory of relativity says about matter compressed to this extent. William McCrea has given the following argument: if we have a sphere of mass M and radius R and we imagine bringing up a small particle of mass m from infinity to the surface of the sphere so as to add to its mass, we can calculate, using Newtonian theory of gravitation, how much potential energy is lost by the particle. Now this potential energy is part of the total energy of the particle and so presumably is related to the mass which is added to the sphere by the mass-energy relation $E = mc^2$, referred to earlier. It therefore turns out that the amount of mass which is added to the sphere is less than was brought to infinity by a certain amount. This amount increases as the radius of the sphere decreases. When the radius reaches the critical length, the extra loss of potential energy exactly cancels out the added mass, so that no mass is added at all. McCrea infers from this calculation that it is not possible for a body to contract inside its critical radius. Any attempt to add matter so as to produce the contraction is doomed to disaster, because the energy corresponding to the mass is radiated away.

However, this view creates severe difficulties when we remember that general relativity does not exist as an all-embracing, self-consistent theory but is only one part of physics. Another most important part is quantum mechanics. Now in quantum mechanics there is a law of conservation of baryons; that is to say, the total number of baryons in any isolated quantum mechanical system is believed to remain constant. The matter brought up from infinity in the experiment just described will consist mainly of baryons. But when it has been brought up and the energy radiated away, the radiation will have the form of light particles. There seems no possibility of avoiding the conclusion that the number of baryons is not conserved, which is in complete conflict with the present formulations of quantum mechanics. Of course, there is no definite inconsistency in either theory; indeed there is no comprehensive theory of general relativity and quantum mechanics.

Now if the number of baryons in the universe has not had some particular fixed value for all time, it must surely be the case that the number of baryons is a certain variable in a suitably comprehensive theory and has its own laws governing its changes. If such a variable is to change, the most favourable conditions for change would surely be the extreme densities which occur in extremely contracting bodies. Moreover, all of this theory has a very direct but somewhat surprising application to cosmological models as a whole. Instead of considering a contracting body one can consider an expanding one; that is to say, one runs the cinema film backwards. Then the sort of arguments we have used lead to rather closely corresponding conclusions in an expanding universe.

Evidence from practical observation

Let us leave all this theoretical discussion for the moment and ask how contraction, catastrophic collapse, and other such phenomena can be observed in practice. At first there was not very much hope of seeing gravitational collapse. People thought that all one could hope to observe would be a neutron star which was almost at the critical point. Now such a star is reckoned to have a radius of only about six miles. If its surface were as hot as the sun, then it would give out about as much light as the earth actually receives from the sun. We would not be able to see such a star in its state before collapse unless it were as near as the nearest known stars. No such object has been found and it would of course have been very good fortune if there were one so close to us.

In recent years, however, a new possibility has been pointed out. It centres on the new science of radio astronomy which we mentioned in the first chapter. In 1960 the source labelled 3C48, that is to say, the forty-eighth one in the third Cambridge catalogue of radio sources, was identified by comparing an accurate radio position of it with the sky survey photographs from Mount Palomar. A faint starlike object with unusual properties was found in exactly the same position. The radiation from the starlike object had a strong ultra-violet component; it was surrounded by a small amount of nebulous matter, and it had a spectrum with sharp lines in it which at first completely defied interpretation. Eventually in 1963 Maarten Schmidt, who was studying another unusual object labelled 3C273, realized that the sharp lines in the spectrum were in fact just the ordinary lines which

As bright as several hundred galaxies, but only a thousandth of their size, quasars are thought to be very distant objects, receding from us at enormous speeds. 3C48 (left) and 3C273 (right) were amongst the first to be identified

one would expect in a gaseous object with the central source of energy. But these lines were shifted to the red end of the spectrum by a very large amount. In fact, if the red-shift is interpreted as a Doppler shift, the recession velocity is about 14 per cent of that of light.

A result like this, predicting that a body could be moving with such an inconceivably large velocity, naturally makes one think that perhaps the red-shift might be accounted for in some other way. As it happens, there is another avenue of explanation worth exploring. We have already referred to the way in which the change in potential energy of a particle contributes to its change of mass, according to Einstein's mass-energy relation $E = mc^2$. If we suppose such a change to refer also to photons, the particles in terms of which light rays are constituted, then a red-shift could be accounted for by a very intense gravitational field. Such an argument is a little hazardous; for, firstly, it mixes the Newtonian idea of potential energy with the special relativity idea of the equivalence of mass and energy. Secondly, and more seriously, it applies this mixture to a particle, the photon, whose rest-mass is zero. However, these objections are all overcome by general relativity. So long as one assumes that

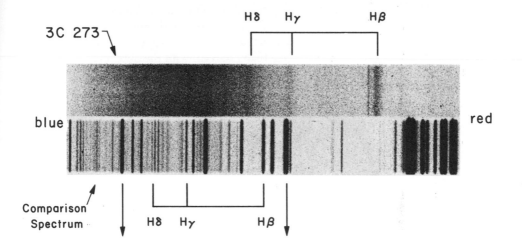

By comparison with the hydrogen emission lines in the laboratory spectrum, the enormous red-shift of 3C273 is apparent (the spectrum has been slightly widened to improve visibility). This 16 per cent shift indicates a distance of 1,000–2,000 million light years

photons have zero rest-mass, their motion in gravitational fields is determined, and a red-shift with change of field from more intense to less intense is predicted. In the particular case of 3C273, Schmidt was able to show, however, that it was impossible to get such sharp spectral lines as this source gave, and at the same time to derive the red-shift not from a recession effect but from gravitational sources.

Acting on the experimental clue, Jesse Greenstein and T. A. Matthews managed to find a correct identification of the spectral lines of 3C48 and discovered that it corresponded to a recession velocity of 30 per cent of that of light. Actually 3C273 had been seen for many years and photographed on about 3,000 different occasions, but without realizing its significance. Now the enormous red-shifts of these quantities suggest, according to the last of the definitions of distance (mentioned in an earlier chapter), that they must be a great distance away. But if they are so far away the amount of energy which is being poured out of them in the form of light, and, even more, in the form of radio energy, must be tremendous. Hoyle suggested that the energies were being produced by an unknown mechanism, using gravitational energy released by gravitational contraction.

After this, discoveries succeeded each other in rapid succession. In 1965 a red-shift was found which corresponded to the frequency of each spectral line being actually halved, with the obvious interpretation of an enormous distance. Many red-shifts of this size are now known. Moreover, the objects vary in intensity as time goes on, both with respect to their radio emission and with respect to the optical emission which is brighter at some times than at others. In 1965 it was realized that there was a very large population of such so-called quasars (for 'quasi-stellar radio source'), some of them radio sources and others only optical sources. For the radio sources it has been established in the past few years that they must have exceedingly small diameters. Most people agree now that the line spectra first observed arise in the hot gas which surrounds, and is probably more extensive than, the actual source. As far as one can tell, the gas is of a fairly normal chemical composition, somewhat similar to that of the sun and the stars in our own galaxy.

The suggestion by Hoyle and William Fowler, that gravitational collapse might be responsible for the energy of these sources, gave a great impetus to the work of the theorists in general relativity. However, the variable nature of the quasars made more people begin to question the nature of the red-shifts. There are really two questions one can ask. Firstly, are the red-shifts due to velocity at all? Secondly, if they are due to velocities, are these velocities those caused by the expansion of the universe and therefore implying a corresponding enormous distance, or are they, as J. Terrell suggested in 1964, due to objects which are being thrown out from the central part of our galaxy at exceedingly high speed? The question of variations in the amount of light and radio signals was really crucial. People felt intuitively that such objects, which were a hundred times more luminous than the nebulae we knew, could not possibly be so small that the amount of radiation from them would vary significantly over a period of years or months. In other words, it is really very difficult to make a consistent

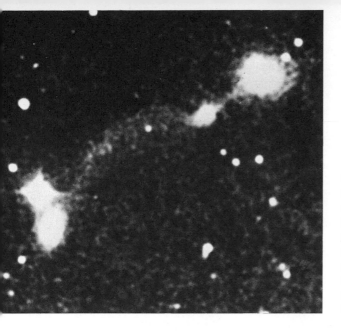

Terrell's supposition that quasars might be objects thrown out at high speed from the centre of our galaxy is lent some colour by similar phenomena elsewhere. This triple galaxy IC 3481 suggests that enormous outbursts of energy can take place in the centres of galaxies with the result that large parts can be ejected with velocities of at least thousands of miles per second

model of a quasar if it is at a cosmological distance. But if it is not at a cosmological distance, then the red-shift cannot be explained simply as due to the expansion of the universe, and it must be associated with the object. Either we return to Terrell's suggestion of very high local speeds or else to some sort of gravitational red-shift. The trouble with Terrell's suggestion is that there should surely be some blue-shifts corresponding to motion towards us as well as the red ones, and none have so far been observed.

The position regarding the gravitational red-shifts is a little better. Greenstein and Schmidt's original arguments, which purported to show that it was impossible for gravitation to explain the red-shift, made the assumption that the observed line spectrum came from a thin shell of gas surrounding a massive object. In 1967, however, Hoyle and Fowler showed that it was possible to overcome this difficulty. If the gas which is giving the line spectrum is concentrated in the centre of a massive object instead of being diffuse, it will be at a region with very low gravitational potential energy and the required red-shift can be produced. However, the object must not be opaque to the radiation, so it could perhaps be made up of a

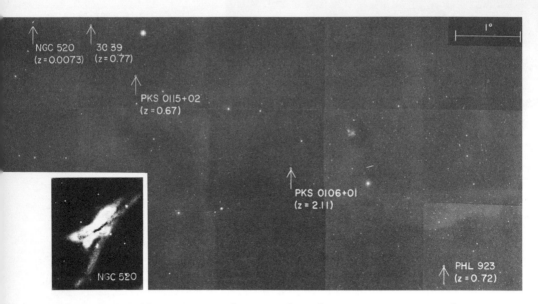

NGC 520
(z = 0.0073)

3C 39
(z = 0.77)

PKS 0115+02
(z = 0.67)

1°

PKS 0106+01
(z = 2.11)

NGC 520

PHL 923
(z = 0.72)

H. C. Arp has looked for correlations between positions and alignments of quasars and of peculiar nebulae. The main plate shows one of his most convincing pieces of evidence: four quasars in a line near the peculiar nebula NGC 520 (which is shown in more detail inset). If such alignments are not due to chance, it suggests that the quasars were thrown out of the nebula

large number of very compact stars, perhaps neutron stars. So far no satisfactory model of this type has been constructed which would be able to explain the gravitational red-shifts.

Meanwhile, a number of observations have come to light which are very difficult to interpret. H. Arp put forward several arguments to support the idea that quasars are not really at cosmological distances at all but are in the galaxy. He looked for correlations between the alignments of such objects and various peculiar nebulae. All these nebulae are known to be at not too great a distance. His work has been disregarded partly because his statistical arguments are difficult to follow; none the less, some of the examples of alignments between nebulae and quasars are very remarkable. There is, for example, an alignment between four quasars and the unusual galaxy NGC 520, which appear to be very much on a straight line. This suggests strongly that there is some connection between objects outside the galaxy which have different red-shifts. This means in turn that, notwithstanding the difficulties in the theory, the most plausible idea is that the quasars are being thrown out from the centre of nebulae. A certain amount of statistical argument seems to have collected to show that the red-shifts of the spectra cease altogether beyond values somewhat greater than 2, and, moreover, that they cluster together. However, the statistical support for this observation is weak and some people believe that the red-shifts can be of any value, at least up to about 2. If there really were pronounced peaks in the distribution of red-shifts of different values it would necessarily imply that the steady-state theory was quite impossible, unless the red-shifts could be explained as due to the actual structure of the galaxies rather than to their position and motion. For if there are definite regularities in the speed with which nebulae are moving away, then there must have been some unusual event at some time which produced this distribution.

The absence of red-shifts greater than about 2 is thought to be a real effect which could either be attributed to expansion, or equally be explained in the steady-state model. Whatever view we may take about these observations, it is evident that at some stage in the history of a nebula or of a large star a very large amount of energy is released. This energy is so great that it cannot be produced by thermo–nuclear means, and it gives rise to a lot of light, infra red radiation, and radio waves. To provide such energy one must imagine a very large mass in a very small volume. The usual view is, as we have said above, that it is the

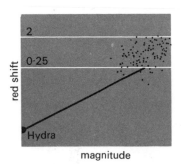

The relation between red-shift and apparent magnitude for some 130 quasars, also showing the line for the normal Hubble relation (see p. 58). The clustering below a red-shift value of 2 is very obvious. Spectra of two of the most distant are shown below. A red-shift (Z_{em}) of 2 is equivalent to a speed of 80 per cent of that of light. The upper spectra are laboratory ones from a helium-argon light, and the black lines crossing the whole spectrum are mercury lines from city lights – a striking instance of the havoc wrought by technology on pure science

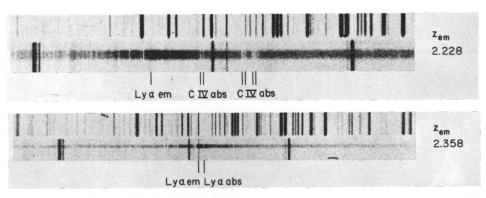

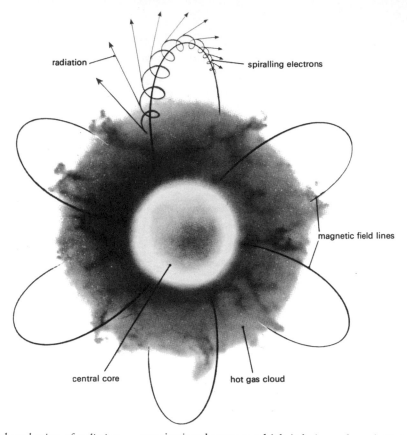

radiation — spiralling electrons

magnetic field lines

central core — hot gas cloud

The actual mechanism of radiation production in quasars is still mysterious. Perhaps there is a comparatively small centre from which a hot plasma emits streams of electrons spiralling in a magnetic field. Such a process certainly will produce radiation but, because of the small size and the high energies needed, such electrons are more likely to lose their energy by another process

gravitational energy which is being released. But that does not specify any mechanism by which this gravitational energy can be tapped. Various suggestions have been made, such as collisions between stars or the gravitational collapse of a single large mass, but none of these models has really met with much favour although an important problem has been uncovered by them. It is always necessary to consider the existence of a very large mass with a size which is not too big, roughly in the middle of such a quasar. Now the densities usually seen in the centre of nebulae are much smaller than these densities. It is not at all obvious to what extent very dense objects could be evolved from situations of very much lower density in the sort of time which we could expect because of the expansion rate of the universe, that is in about 10,000 million years. This is a strong argument against an expansion theory. One could argue in the steady-state theory that the regions where the quasars have these very large masses are the regions of creation in a steady-state universe. There is, however, one way out for those who want to consider the universe as expanding from an initial state. One could imagine a universe which expands some of the time and contracts some of the time; then these regions could have been left over in some way from an earlier contracting phase.

Sources of radiation

As can be seen, the situation is very untidy. Despite all the problems, most astronomers still think that the interpretation of the quasars as being at cosmological distances is the more plausible one. If one takes this view, one is led to believe, because of the relation between the apparent magnitude and the red-shifts, that the universe is indeed an evolving one of the Lemaître kind. There remains a certain number of theorists who take very seriously the alternative possibility, that the quasars are not at cosmological distances at all but are quite close to our own galaxy. This would mean that the extreme red-shift of their light is a property which will be interesting for studying the bodies themselves but will have no relevance for the difficult choice between rival cosmological theories. Rather, they will be of great importance for testing theories of gravitation, since such explosive happenings always occur in regions in which very strong gravitational fields play an important part.

The decade of the 'sixties threw up other striking experimental evidence for cosmology, however, and the balance of this tends to be strongly against the steady-state theory. Apart from observations of the recession rate which might indicate a disagreement with the steady-state model, the most direct source of evidence is the temperature of the initial radiation thrown out by a 'creation' explosion. This radiation will have been cooled by the expansion to about $3°$ Kelvin.

The oscillating universe theory proposes an initial fireball which expands over some 45,000 million years to a maximum size, then contracts to reform the fireball. The cycle then starts again

As a matter of historical fact people began to look for this 3°K radiation on the basis of an evolutionary universe proposed by Dicke, who envisaged that the expanding universe would begin to contract, and so eventually collapse to a glowing fireball of hydrogen. This extremely dense matter would then blow apart again and the cycle would be repeated. The 3°K would then be radiation left by the initial explosion. This radiation is just what the astronomers have detected. The way this came about was that, in the early 'sixties, two scientists, A. A. Penzias and R. W. Wilson, at the Bell Telephone laboratories undertook an experimental programme for finding micro-wave noise at 7 cm. wavelength due to the atmosphere. The apparatus was an aerial which could be pointed in different directions, and it was found that in different directions there were different amounts of noise, as would be expected because of the different amounts of atmosphere. Rather surprisingly, however, when the results were plotted and the extension of the graph was surmised towards the state where the atmosphere contributed nothing, there was a small amount of residual radiation.

If one thinks of the universe in thermodynamical terms one might ask: what sort of radiation is there due to its temperature? To give this some intelligible meaning we must mention a few well-known facts about radiation. Everyone has observed how a poker put into a hot fire emits a red glow. Moreover, the hotter the fire the more the colour of the glow moves from a dull red to a lighter red and so eventually to a white colour. The radiation, in this case light, is described adequately by Maxwell's electromagnetic theory. But this theory makes no mention of temperature, and is quite unable to answer the question of how the colour of the poker varies with temperature. Instead of a poker it is more convenient to think of a closed box, with a small hole cut in one side so that one can peer into the cavity. In this case one can speak of cavity radiation or black-body radiation. If one raises the box to a certain temperature, say, by heating its outside with a bunsen burner, when one looks through the hole one will see radiation with a certain definite spectrum. That is to say, there will be a certain intensity of red light, some of blue, some violet and so on. The energy of the different colours in the light will be determined by the temperature.

One can reduce the problem to a more reasonable form by first drawing a curve giving the intensity of the light against the colour, the colour being measured by the frequency of the light. Then when one has plotted this curve at one temperature, one can deduce it at other temperatures. This was proved by Wilhelm Wien as early as 1893 by thermodynamical arguments. But what Wien was unable to do was to predict the distribution of energy against the temperature at any one particular temperature. The process of deducing the distribution at other temperatures could not even start. Both Lord Rayleigh and Wien himself had used further arguments for this problem, which gave definite but wrong answers. In 1899 Max Planck made a slight modification to these answers and so was able to find the correct value for the distribution of energy against frequency at a given temperature.

In simple terms, then, what is the overall temperature of the universe? As the universe seems to have no connection with what is in some sense 'outside it', the natural answer seems to be that its temperature would be almost at the absolute zero on average.

Actually, George Gamow and his associates had already calculated the existence of such black-body radiation of a few degrees above zero fourteen years earlier, but their paper had been ignored. It seems certain that on the basis of an evolutionary theory the deviation of the distribution of this radiation from black-body radiation should be exceedingly small, perhaps about one part in a million. To begin with, the various points on the graph of the radiation intensity were all in good agreement with black-body radiation of a temperature of $2 \cdot 7°$, exactly as expected on the evolutionary hypothesis. More recently, however, more points on the graph have been found and

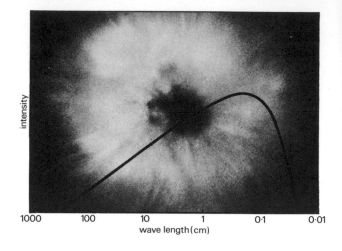

If the background radiation is the debris of the initial explosion, it should be very close to the graph plotted here

some of these seem to deviate considerably from the black-body graph.

However, this radiation could possibly be explained in terms of the continuous creation process in the steady-state theory, and, if it is indeed not black-body radiation, it would be the obvious explanation. If such an explanation could be provided it would carry with it a valuable extension of this theory in showing the nature of the creation process. But the most recent observations show to what a high degree of accuracy this radiation is the same in whatever direction we look. This is, of course, just what we would expect if it is the residue of the original big bang, but if it is produced as a background in a creation process, the number of centres at which creation is taking place must considerably exceed the number of nebulae. While this is not impossible, it is certainly hard to believe; this means that the pressure to believe in a big bang, of which this radiation is the debris, is all the greater.

Pulsars

Whatever may be the ultimate conclusion about the significance of quasars and their structure, they have given rise in the 'sixties to another discovery of remarkable importance in astrophysics. In the investigation of quasars a great deal of importance has been attached to the fact that they are very small in size. One of the ways we know this is that the radiation from them passes through the hydrogen clouds sent off by the sun and this causes the radio waves to fluctuate. This is just the same as the way in which our atmosphere causes the light from stars to twinkle. Ordinary nebulae do not twinkle, because they are too large, but quasars do, and accordingly, one of the routine ways of detecting them is to search for radio sources with fast fluctuation.

In an investigation of this kind in Cambridge in 1968, led by Anthony Hewish, a most unusual radio source was discovered by his assistant Miss S. J. Bell. In analyzing the charts for fluctuations of the kind expected, she found what was at first thought to be interference from a police transmitter; but the signals reappeared about once a week, vanished for about a month and then reappeared again as unexpectedly as before. They were a number of sharp flashes of radio emission only about a hundredth of a second long occurring at exactly maintained intervals of a second, showing that the body must be very small. For if a body of the size of the sun were suddenly switched off, the times taken by light from different portions of it to reach an observer would vary so much that the sharpness of the switch-off would be blurred. Indeed, one can say that these transmitting bodies, the so-called pulsars, must be no bigger than the earth. Moreover, such a body must be well outside the solar system not to be observed optically. Some dozens of pulsars have now been found and they all have extremely similar properties. They cannot be planets of stars other than the sun because their orbital motion gives rise to no Doppler shift. One can conclude from this that they are not artificial signals, a possibility which is not so

Pulsars are characterized by the emission of sharp pulses of radio energy, occurring at regular intervals of about one second or less

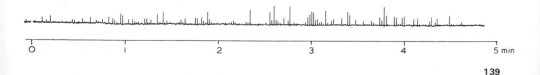

remote as at first appears. Consider how many stars there are in the universe of roughly the size of the sun, and how many of these might hold planets at a distance comparable to the earth's and thus capable of supporting a form of life similar to our own; the number turns out to be immense so that the chances of there being some life somewhere in the universe are very good indeed. Such life, somewhat like ours, might be a long way behind or a long way ahead of us in development. Some of the planets, perhaps half, where life has occurred are very likely to contain beings who have discovered electromagnetic radiation. These might therefore be deliberately transmitting signals to try to find whether there was anyone else in the universe with them. Again, such signals might be the normal radio transmissions of the planets themselves.

If, however, there is no evidence of planetary motion, then the signals must be coming from something like a star producing its own energy, and it becomes inconceivable that any form of life should be responsible for the signals. Radio signals on different wavelengths are affected by the ionized inter-stellar hydrogen. Differences in wavelength of the pulses gives them different times of arrival and from this it is possible to work out the distances of the pulsars. They turn out to be something of the order of a hundred light-years, and so well within our own galaxy. As more pulsars came to be discovered, the most remarkable fact about them was found to be how very well

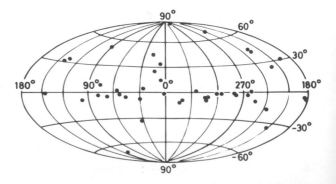

The distribution of known pulsars relative to our galaxy, showing a certain amount of clustering in the galactic plane

they keep time. In fact their time-keeping was accurate to about one part in a thousand million over a period of weeks. Further investigation showed that there was a systematic decrease of the repetition frequency of the pulses, and this of course must be an important factor in understanding their physical nature.

Perhaps one of the most peculiar facts about the pulsars is that optical astronomers have been almost completely unable to correlate them with light sources. Now that several dozen pulsars are known, and correlations of this sort have not succeeded, the fact must be of some significance. This significance will not be greatly diminished if in the future some correlations between pulsars and optical sources are found. Most of the pulsars found are in the plane of the galaxy, but by no means all of them. This recalls the situation with the original radio sources when there was a conflict between whether they should be associated with the galaxy or not. Perhaps there are two families of pulsars.

The largest fully steerable instrument in the world, the 250-foot Mark 1 radio telescope at Jodrell Bank has provided us with much of our information about the pulsars in the northern sky

A number of theories have been put forward about the production of pulsars. One is the theory which we described earlier about a star at the end of its life. We should perhaps at this point go into a little more detail about the theoretical views on the life-cycle of a star. The usual thermo-nuclear interaction which produces the radiation from stars is the turning of hydrogen into helium. The hydrogen then collapses inwards, under the influence of gravity, but a star of a stable size, such as the sun, is produced by a balance between the pressure of radiation outwards and the gravitational force inwards. When all the hydrogen has been converted into helium, the helium in turn takes part in another interaction into higher elements. Eventually, however, all the possible atomic interactions are exhausted and an end-point is reached.

Now there are various possible situations at such an end-point. In the first place the star may finish with a mass rather less than a critical mass, which can be worked out. This mass is slightly bigger than that of the sun, and such a star is known as a white dwarf; its volume is about that of the earth and it gradually cools. If the mass is a little greater than this critical mass, then the core of the star collapses and, as was explained above, the electrons and protons combine together to form neutrons. The middle collapses, the resultant energy bursts out, and one has the phenomenon known as a supernova in which the outer matter is sent off, leaving in the centre a neutron star. If we have more than ten times the mass of the sun, there is no stable configuration at all and the star appears to collapse to infinite density. This is the final catastrophic collapse mentioned earlier. Supernovae do indeed occur; one is observed every 300 years or so, and in fact one is due any time now.

The existence of pulsars seems, however, to imply the existence of neutron stars and so it seems very likely that the state of complete collapse must also exist. How, then, can we hope to explain the observed behaviour of pulsars in terms of neutron stars? The only hope seems to be what might be called a light-

In 1937 a photograph of a far-distant galaxy revealed a supernova. Over a year later, even after a 45-minute exposure, it was faint. In 1942, with the long exposure magnifying the surrounding stars, the supernova had disappeared

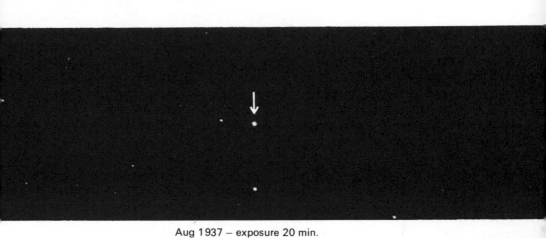

Aug 1937 – exposure 20 min.

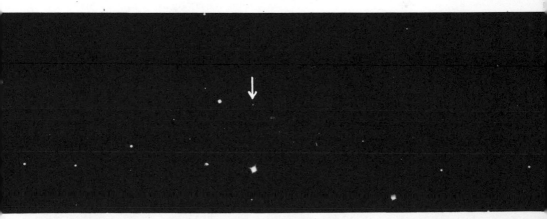

Nov 1938 – exposure 45 min.

Jan 1942 – exposure 85 min.

house theory. In this the regularly recurring pulses are thought to be caused by a point on the star radiating and the star as a whole rotating at a high angular velocity, so that the beam of radiation from the point sweeps across the earth's observatories at regular intervals. One might have thought of the white dwarfs as possible candidates for a lighthouse theory, but now that American scientists have discovered a pulsar with a pulse rate of 30 per second the theory is refuted, for white dwarfs could not spin as fast as that without flying apart.

This exceedingly rapid pulsar is in the Crab nebula, mentioned before (p. 21) as the remnant of the supernova explosion observed by Chinese astronomers in 1054. It is noteworthy that the central star of the Crab nebula is flashing with the same period as this pulsar. A neutron star itself cannot be radiating, because for the radiation to overcome the enormous gravitational force the star would have to be hotter than is acceptable. There must therefore be some other detailed

Photographed during (left) and between (right) its extremely regular flashes of light, this star in the Crab nebula has a period and position coinciding almost exactly with that of a known pulsar

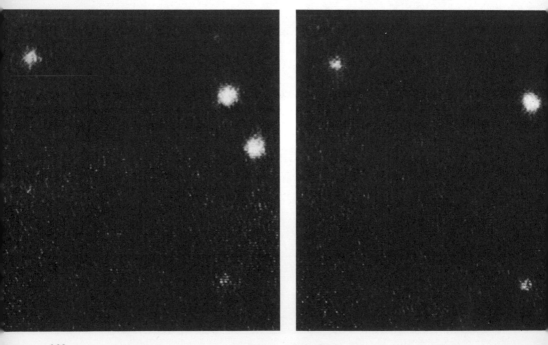

mechanism for producing the radiation. The most popular theory is of some sort of dynamo effect. Ordinary stars usually have magnetic fields. Since neutron stars have become extremely compressed compared with ordinary stars, it is likely, because of their high speed of rotation, that they will have very intense magnetic fields indeed. While it is still unclear what mechanism should be chosen for the pulsars, there is no doubt that we shall soon learn a great deal more about stellar constitution. Moreover, if the speculations of Harrison in 1970 are correct, a study of pulsars will be of great value in checking the theory of relativity. Harrison supposes that the catastrophic collapse of a star is more likely to result in what he calls a bouncing core configuration, rather than a neutron star. This is a situation in which the catastrophic collapse in the centre gives rise in some way to a bounce and the resultant radiation pressure at the centre prevents the collapse of the remainder of the star.

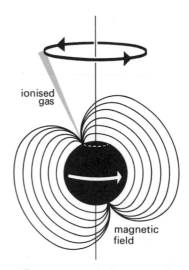

There are several theories on the mechanism of pulsars. One suggests that plasma (ionised gas) streams out along the magnetic field lines, emitting radio waves at right angles. The rotation of the star causes a 'lighthouse' effect – we receive pulses only when the beam is swinging towards us

Inconclusive evidence

The intensive continuing study of the quasars and pulsars is to be expected to lead to considerably more information about the nature of the bodies that exist in the universe and about their distribution. Any information of this kind is likely to be most useful in deciding between the rival cosmological theories. Although nothing definite can be said at the moment, the balance of opinion is heavily against the steady-state theory.

It should be mentioned that Hoyle, despite his earlier doubts, is still very interested in the steady-state theory, and in 1968–69 he and J. V. Narlikar again took up the arrow of time argument discussed earlier (p. 114). Their investigation is into the relationship between cosmology and quantum electrodynamics (that is, the quantum theory of the electromagnetic field). They believe that a decision between rival cosmological theories could be reached by means of experiments carried out by atomic physicists. Essentially, their argument is that there is a strong connection between the large-scale structure of the universe and the electrodynamics of particles. Thus they are taking Mach's principle to its logical extreme. They claim that the failure to take note of this connection in the past has led quantum electrodynamics into calculations involving infinite quantities. It is true that methods have been found for removing the infinite quantities again, but the result can hardly be considered satisfactory. None the less, workers in the field were confident in the predictions of quantum electrodynamics. One of the reasons for this confidence was the extremely high accuracy with which the results of certain experiments could be predicted. In particular, the Lamb shift (a slight displacement of the lines in the hydrogen spectrum) was predicted very accurately. Hoyle and Narlikar were able to show that an alternative explanation of the Lamb shift was possible, in terms of the response of the whole universe to the particle. In such an alternative explanation there were no infinite quantities. But – and here lies the relevance of their work – they were able to show that this calculation is possible with reference to the steady-state theory only. Accordingly they consider that their calculation makes the steady-state model extremely probable; and they regard such experimental evidence as more valuable, because it is more accurate, than the astronomical data which seem to favour expansion theories.

On the other hand, there tends to be a built-in irrational prejudice in favour of 'direct' astronomical observations and against a 'theoretical' argument like that of Hoyle and Narlikar. Such a prejudice ignores the fact that all experimental results have meaning as part of a theory. Evidently, in order to make a final decision, we shall have to have some new observations and these are likely to be different in kind from the ones we have already. In the next chapter we shall try to survey the possible directions from which these will come.

We have now come quite a long way from our starting point. We have tried to justify, in terms of a theoretical picture of space and time, the explanation of the brute facts of experiment, but a bewildering amount of uncertainty remains about the nature of the objects and their positions and motion in space. All that can be said with certainty is that cosmology is developing very fast, and that, with any luck, in the next few years there will be much more information than at present. This chapter is a survey of the new kinds of information and the effects it may be expected to have on our theories. This will necessarily be a somewhat fragmentary description since none of the developments to be described is in a fully developed state at present.

In the immediate future there are several fields calling for more detailed study. The radio counts will, of course, continue. There is a strong suspicion at present that they give substantially different answers at different frequencies. If this is so, it is evidently of considerable importance. Also of importance are recent claims that at the very greatest distances there are fewer radio sources. Such great distances correspond to long ago 'before the galaxies were condensed' and strict confirmation of this would be strong evidence for an evolutionary theory. Again, the red-shifts of some quasars are so large that it is difficult to accept them as due only to recession. Any further evidence of this would be important. Many more measurements need to be made, too, of the 3°K background radiation. Some of the later observations seem to have been less in conformity with a black-body curve; if the radiation is to be attributed to the initial big bang, the general opinion is that it should be strictly black-body radiation, correct to about one part in a million. Finally, amongst the problems of the immediate future, there is good reason to expect that there are large scale and significant irregularities in the distribution of matter in the actual universe. Such irregularities could be studied by a more detailed analysis of the angular distribution of quasars and of the 3°K radiation.

We cannot claim to know much about what measurements are likely to be performed even in two or three years', let alone ten years' time, but in the next decade we expect only to hear of greater refinement in the performance of observations already made, not of anything significantly new. However, there might be a surprise in store for us. The experiments described in the present chapter seem to bc of a more forward-looking kind which may be expected to become much more important. And equally, here we may be very disappointed!

To extend our understanding of the universe, we must increase the number and accuracy of observations. Long baseline interferometry is a technique used to place objects accurately by combining signals from widely separated radio telescopes. The longest baseline used so far is between Algonquin Observatory (right) in Penticton, Canada and Parkes Observatory (opposite) in Australia – 7,500 miles

Before describing the experiments it is necessary to say a little more about the recurrent problem of distinguishing between the evolutionary and the steady-state universe. The steady-state universe is understood fairly well but it is time now to consider the further details of the evolutionary universe. In a very general sense the expanding universe was described earlier, but the increasing knowledge in atomic physics has lent this theory considerably more precision. It is now much more of a laboratory theory than it was when Friedmann and Lemaître originally put it forward. The following description – a somewhat speculative one – will be given without any of the reasons lying behind the numerical estimates. Most of these estimates are made with an extensive knowledge of constants arising in atomic physics.

A comparison of theories

It is suggested that the expanding universe originally starts from a state which we cannot describe except by saying that it is of infinite density. After about one millionth of a second it reaches a density of the order of that of an atomic nucleus. The temperature at this stage is extremely high, ten million million degrees. Of course in such extraordinary circumstances the composition of matter is quite different from ordinary matter. Here is the difficulty which the steady-state theory proposes to overcome by not admitting such unusual circumstances. If, instead of taking this logical standpoint, one hopes for the best and imagines that the laws of physics, derived in the ordinary circumstances on the earth, do still apply in such different circumstances one can conclude that protons, the stable constituents of matter, will be unstable at these temperatures and densities and undergo reactions, so that they are converted into other particles. Anti-protons and anti-neutrons are created and there is a high energy density of neutrinos. The universe begins to expand and so the matter and the radiation cool down. When the temperature drops to ten thousand million degrees most of the unusual particles, other than the positrons, have disappeared. Only the photons and electrons, positrons, neutrons and neutrinos are left. By the time the temperature falls to a thousand million degrees most of the electrons and positrons are annihilated and the energy of radiation is correspondingly greater.

From here onwards the neutrinos are irrelevant to the further stages of the evolution. Between a temperature of a thousand million degrees and about a hundred million degrees the elements have begun to build up from the positrons and neutrons. At the higher temperatures the nuclei are unstable. They will be formed and then they will break up again. But as the temperature falls the nuclear reactions become too slow compared with the expansion of the universe and the temperature range mentioned is about right for the build-up of elements. The time that the universe spends in this range is something in the order of a thousand seconds. During this time the helium begins to be formed in various isotopes, and by the end of the first half-hour 20–30 per cent of the protons will have been converted into helium. As the temperature falls further the electrons and ions will have re-combined, leaving very little matter frozen in the ionized state – perhaps a thousandth or less of all the matter. So long as the radiation density is too great compared with the matter density it is impossible for gravitational condensation of gas masses to take place, so that galaxies and stars cannot be formed. This is because the radiation can be thought of as a gas whose particles have no rest mass, and so its pressure is too great for its own gravitational attraction to hold it together.

However, the expansion causes the energy density to decrease steadily so that the radiation energy eventually diminishes to room temperature. At about this time globular clusters with a mass of perhaps a hundred thousand solar masses begin to be formed. These will be formed before the galaxies, and once these have been formed stars begin to arise. The observed mass of stars ranges from about a tenth of the mass of the sun to sixty times that mass. Since the life-span depends on the amount of mass which can be consumed and upon the luminosity, and moreover the luminosity is proportional to the third power of the mass, it follows that the life-span of one of the largest heavier stars is of the order of some twenty thousand million years whereas one of the lightest stars will have a life-span of about a million years. Now according to evolutionary cosmology the age of the universe is something of the order of ten thousand million years so that some of the oldest stars may contain a certain amount of the matter manufactured in the first half-hour. Such stars ought then to contain nearly 20 per cent helium but none of the heavier elements like iron and carbon. As a matter of fact some stars are known for which the heavy element content is much less than 1 per cent and in certain cases there are no spectral lines at all from heavy elements.

The situation is however rather disappointing with regard to another possible experimental check on the theory. According to what has just been said the amount of helium present in the old stars should be something like 20 per cent. Unfortunately these old stars have low surface temperatures, of the order of 3,000°, and helium lines can only be excited in the hottest stars with a surface temperature of 10,000° or more. Accordingly we cannot hope to determine the amount of helium present in such stars. There are one or two of these stars which have evolved in a different way from the standard ones and they have a higher temperature and have helium lines present in them, but the exact helium content is not known. Further experiments are evidently needed here.

Neutrino astronomy

It will probably be clear from this description that another field which may prove very rewarding in cosmology is that of neutrino astronomy. Many atomic nuclei undergo the interaction known as β-decay. This is a nuclear transformation which is accompanied by the emission of an electron as already discussed. Since its original prediction by Pauli many experiments have given independent evidence of the neutrino and its properties, though such evidence is rather hard to come by, since the neutrino is electrically neutral, has zero mass and a spin which is like that of the electron. The energy-releasing processes which go on in the stars release neutrinos. Accordingly, if we study the flux of neutrinos from the stars we ought to be able to find something out about those processes. Unfortunately neutrinos interact only rarely with other particles, which makes them very difficult to detect. However this fact can be turned to advantage. In astronomy we are always entirely dependent on observation rather than experiment. We have to sit and wait for the electromagnetic radiation (light or radio waves) to enter the solar system and then we must detect it by instruments on the ground, or at best in a space observatory. But these quanta of radia-

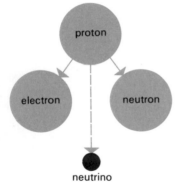

β-decay of the atomic nucleus results in the formation of an electron, a proton and a neutrino

Of all the particles produced by the atomic processes we think are occurring in the sun, only the neutrino can escape into space. These particles, which can tell us more about the sun, have no mass, no charge and travel at the speed of light. The only way to capture them is in giant fluid-filled tanks buried deep underground (opposite)

Clouds of obscuring dust, such as the Horsehead nebula in Orion, absorb radiation from stars beyond them and retransmit it in a changed form, thus making their study extremely difficult

Opposite: the neutrino leaves no track in the bubble chamber and only by its rare interaction with other particles can its presence be detected. Here, one enters at the bottom of the picture and interacts with another particle to give a proton (short track on right), a muon (long central track) and a positron (tight spiral)

tion which form the light and the radio signals from outer space interact very strongly with matter. Since we first have the signals passing through the atmosphere and then through our apparatus we cannot be detecting the photons which were originally released from the stars. Indeed the energy leaves the outside of the stars in the form of low-energy photons, whereas we have good reason to believe that it is originally produced in the interior in much higher energy forms. Even when higher-energy photons are emitted freely from an object, as can happen sometimes, the radiation is likely to be absorbed or scattered before it reaches us. Moreover the inter-stellar or inter-galactic gas which lies between us and the object re-transmits another form of radiation with the same energy. Similarly the magnetic field in the inter-stellar gas may deflect any rays which contain charged particles. None of these obstacles applies in the case of the neutrino.

On the other hand the very weak interaction between the neutrino and matter makes it difficult to detect and an absorber for neutrinos must be very extensive to detect a significant number. There is the

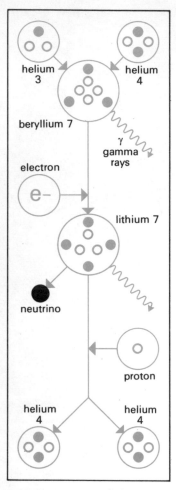

helium
3

helium
4

beryllium 7

γ
gamma
rays

electron

e−

lithium 7

neutrino

proton

helium
4

helium
4

*One of the energy-producing proces-
ses in stars*

additional disadvantage that only a small fraction of the energy that stars emit is emitted in the form of neutrinos. Nonetheless as considerable interest is being shown in neutrino astronomy the reasons for this interest and what is expected from it may now be stated in further detail. We know very well the general mechanism of generation of energy in stars. They are thermo-nuclear processes some of which we are able to imitate on the earth. But we have come by this knowledge largely through a complicated process of inference and elimination and we have little direct observational confirmation of the theory. However, the nuclear reactions concerned are some in which neutrinos are emitted, and if these neutrinos could be observed we would have direct observational confirmation. Indeed, in looking for neutrinos from the sun, since this is the nearest star, it turns out that one of the reactions which produces them is very sensitive to the temperature. Here by temperature we mean the temperature at the centre of the sun, which is the important one. Thus if we could measure the proportion of neutrinos produced in this reaction even with an error of as great as 50 per cent, we should know the central temperature of the sun to about 10 per cent. This would be a great increase in our direct experimental knowledge of the interior constitution of stars.

Of course, because of the general production of a few per cent of neutrinos in all the stellar interactions, there must be a considerable flux of neutrinos in the universe arising from the stars in general. But such a background flux will be difficult to detect for a number of reasons, so it is more promising at the moment to look for discrete sources of neutrinos, preferably inside the solar system. The statement made above, that in normal stars only a small fraction of the energy will be emitted in the form of neutrinos, needs a certain amount of qualification. In some periods in the evolution of the star the neutrino emission may for a time become a major fraction of the whole. For example, Fowler has estimated that the neutrino energy which stars give out after they have finished burning their hydrogen is about one ten thousandth of their total rest mass. Such a proportion is of the same order of magnitude as the total energy emitted in the hydrogen burning. But the stars are near the end of their life by this time so that the relative luminosity in terms of neutrinos is very great. This luminosity becomes greatest just before the final catastrophic collapse and is probably what we see in supernovae outbursts.

It is appropriate here to consider in a little more detail the whole history of stellar evolution. After the hydrogen has been converted into helium by what is now a commonplace thermo-interaction the next stage is that of burning up the helium. This gives rise either to carbon or to oxygen, depending on the mass of the star. The helium-burning reactions do not emit neutrinos. Carbon burning which comes next gives rise to nuclei of nitrogen and oxygen, and oxygen burning in the other stars produces nuclei of a somewhat similar kind. Again there is very little emission of neutrinos. However, on the other hand, these reactions go on at temperatures high enough for pair annihilation to take place. This can take place in

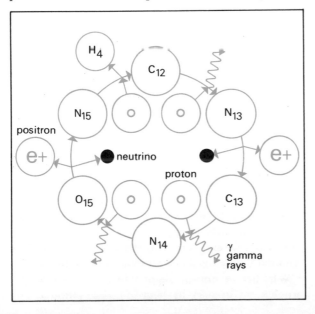

Carbon-nitrogen cycles also occur in the stars, giving rise to helium nuclei, neutrinos and high-energy gamma rays. The positrons eventually collide with electrons, annihilating each other

two ways, either by giving off high-energy radiation or by giving rise to a neutrino and an anti-neutrino. In the second case the neutrinos can directly escape from the star and it seems likely, although there is no direct experimental confirmation as yet, that this process will operate in stars. If it does so, then neutrino pair production becomes more and more important when the temperature begins to exceed about 1,000 million degrees. The later stages in the history of the star consist of other nuclear interactions which occur faster and faster, building up the various isotopes beyond silicon and finishing with nuclei of the iron group with atomic weights of between 50 and 60. When the central region of the star has reached this situation there will be a collapse and a supernova explosion as described before. Direct observation of neutrino flux might well help to establish proper experimental confirmation for these theoretical predictions.

There is another way in which neutrino astronomy may become important. It is exceedingly difficult, as we said just now, to detect the background flux. It might happen however that the background radiation of neutrinos is much higher than the estimates suppose, and in that case it might be possible to detect it. It would then be natural to ask where this background radiation comes from. The general inference in the case of any large background radiations is usually that they must have arisen in the course of the early evolution of the universe when circumstances were very different from what they are now. This is exactly in line with the corresponding situation with electromagnetic radiation. If this inference is correct, observations of this kind provide strong evidence in favour of an evolving universe rather than a steady-state theory. However, as was stated in connection with the radio background, the steady-state theory is not altogether lost because such an appearance of energy might be caused by the creation of matter at all parts which is required by the theory of the steady-state universe.

Whether or not any light will be thrown on the problem of quasars by neutrino observation is unfortunately somewhat in doubt. There are various ways in which quasars might produce their energy, and since there is no evidence of which is the correct one it is impossible to tell whether one particular mechanism, which would produce a good number of neutrinos, is likely to be the favoured one, or another which would produce none. Accordingly the amount of information we can expect to get about quasars from neutrino astronomy is at present somewhat limited.

X-ray astronomy

Another field of observation which may, however, prove extremely important to cosmology in the future is that of X-ray astronomy. X-ray astronomy is the study of objects through the X-rays which are emitted by them. Since only extremely energetic interactions produce X-rays, it is necessarily the study of rather high-energy processes in the universe. One could see a vague beginning for X-ray astronomy in the experiments of Edlen, who showed the character of the solar corona which until the 1880's was a complete mystery. During an eclipse one can see a region of light and flares around the sun but there was doubt in the last century as to whether this was a real physical effect or an optical illusion produced by the moon's light. By the end of the last century the light from the corona was measured by the spectroscope and definite spectral lines were recognized. This showed that the corona was real, but raised a new problem, because these spectral lines did not appear to correspond to any of the known emission lines. Edlen identified most of the mysterious lines. The strongest ones proved to be emissions from iron but not iron in a normal state. They were from iron which had lost either 9 or 13 electrons. It turned out that the corona was a gaseous envelope at a temperature of about a million degrees surrounding a rather cooler sun; iron is able to exist in the electron–deficient form described only in a medium of this type.

Although Edlen's discovery was made in the 'twenties, X-ray astronomy did not really begin as an

experimental study until 1949. The difficulty was that, although X-rays are normally thought of as penetrating radiations, they are stopped by our atmosphere and it is necessary to have observations made from above the atmosphere. The lower-energy X-rays produced by the sun's corona, for example, are stopped at about 60 miles above the earth. V2 rockets, confiscated after World War II, were used in the first experiments. Since then a considerable study of solar X-rays has been made. However, as the corona would in any case have suggested the existence of these, this would not by itself have given rise to the present excitement about X-ray astronomy. The discovery of other X-ray stars was quite unexpected. Indeed in the late 'fifties scans of the whole sky were made for X-ray stars, using rather insensitive equipment, but none were found. And then in 1962, during an unsuccessful search for X-rays from the moon a strong source of X-rays was found to come from the direction of the centre of the galaxy.

X-rays, such as those produced by the sun's corona (opposite), cannot penetrate the earth's atmosphere. Detector instruments, at first carried by V2 rockets (above) are now carried by radio-astronomy satellites

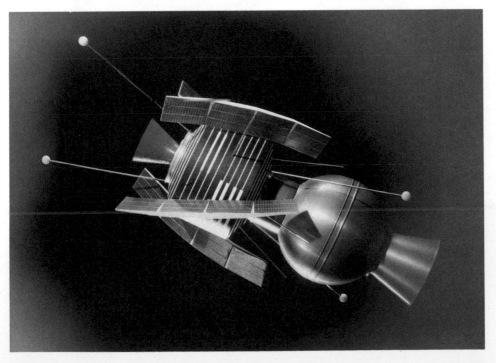

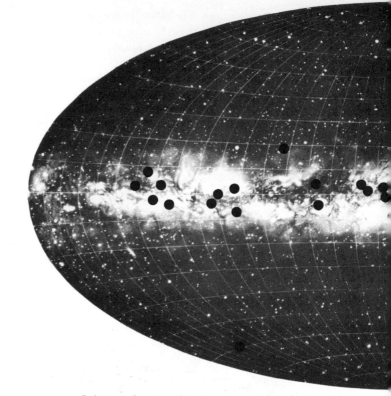

The X-ray sources known at present, superimposed on the Milky Way. Only a few lie off the galactic plane, so most of them probably belong to the galaxy. At the extreme right is the source in the Crab nebula, and on the centre line, just above the Milky Way, is the strong source in Scorpio

It is now known that there was some error of position and the real situation is even more peculiar than was at first thought. The main source of these X-rays striking the earth is not the centre of the galaxy but an X-ray star in the constellation of Scorpio. Its position is known to within about one degree but, most oddly, there is no unusual visible object or radio star in this position. Yet the energy of its X-ray emission on reaching the earth is equivalent to the amount of energy which one would receive as visible light from an extremely bright star indeed.

Since 1962 nearly all the sky has been scanned for X-ray stars, and about 40 of them have been found, the majority in the general direction of the centre of the galaxy but none of them comparable in strength to this original source. Most of them lie in the plane of the galaxy and only two of them are as far as 90° away from

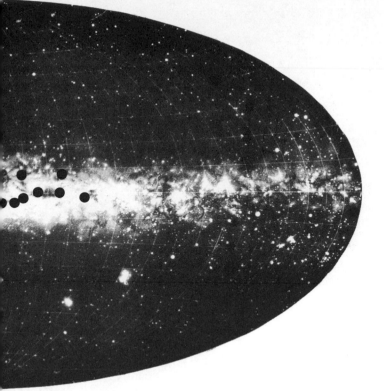

the galactic centre. This distribution is probably roughly of the same kind as the actual visible stars in the galaxy.

It seems as if there are two types of X-ray star, apart from the sun. One of the X-ray stars, that is the one which is farthest from the centre of the galaxy, has now been definitely identified with a known astronomical object, the Crab nebula, which was mentioned earlier. This is the third strongest star and it is visible with good telescopes as a peculiar optical object. Both the optical and the radio signals from this part of the sky are thought to be produced by vast clouds of high energy electrons trapped in a tangled magnetic field, producing a kind of radiation known as synchrotron emission, because it is observed inside the synchroton electron accelerator. The general belief is that the X-rays in this object are produced in much the same way.

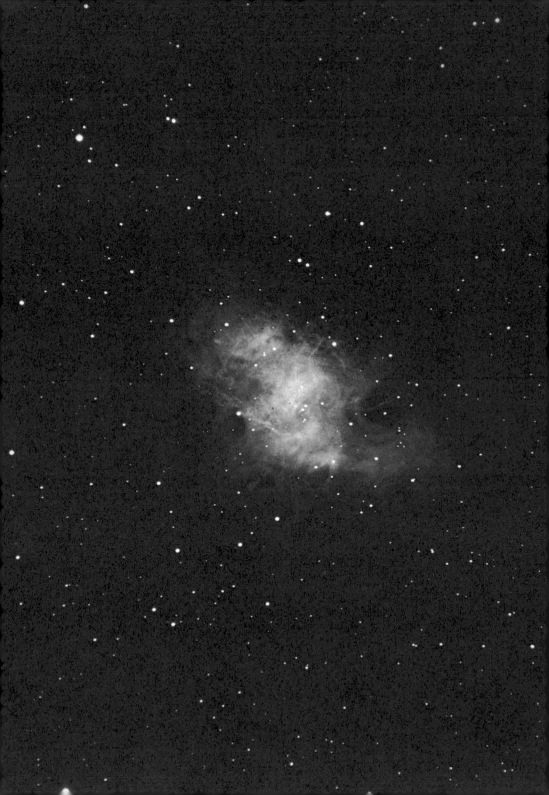

There are still puzzles about this source however. It was remarked earlier that the Crab nebula was observed as an explosion by the Chinese astronomers in 1054. Now the life-time of electrons which have sufficient energy to cause synchrotron radiation is very much less than the time which has elapsed since 1054. The observed emission therefore raises the problems of how these high-energy electrons have been replaced, and in what form the energy is stored up.

From the point of view of cosmology, a more interesting problem than the study of X-ray stars would be the analysis of the X-ray background. It seems probable on general grounds, though experimentally not quite certain, that the background comes from outside the galaxy. One view which could be taken is that it is simply the unresolved sum of all the extra-galactic sources, that is, stars in other nebulae. But another alternative is that it could be radiation arriving from the most distant parts of the universe because it was created in the initial explosion. If this could be established, that is if one could be sure that the majority of the background radiation came from the most distant parts, the steady-state theory would be unable to account for such a situation and this would point strongly to an evolutionary universe.

Gravitation

The next topic, which is among those certain to be important in the next decade, is connected with the question of general relativity. We know that general relativity may be an exceedingly useful tool for describing the world on a cosmological scale, although the steady-state theory suggests otherwise. Accordingly, everything that we know about general relativity is bound to be useful in making a decision on the cosmological problem. We mentioned earlier how difficult it is to distinguish between the predictions of general relativity and those of Newtonian gravitation except in certain special circumstances. The theory suffered considerably in its early years from the fact that it was used to describe just those systems which

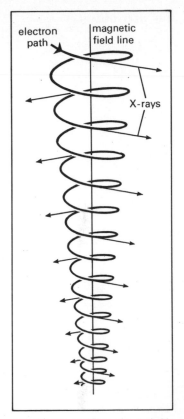

Synchrotron emission, the probable cause of X-rays. As the electrons spiral in the combined electric and magnetic field, they lose energy, and the radiation changes from X-rays to radio frequency

Opposite: the remains of a giant stellar explosion, the Crab nebula is remarkable in being a source of both radio-waves and X-rays. Electrons are accelerated to high energies in the yellow centre, and hydrogen at a temperature of about 40,000 degrees causes the red colour of the outer filaments

were already very well described by Newtonian mechanics, for example the motion of the planets round the sun. It was only after World War II that attention was turned to finding those problems for which Newtonian gravitation would be an exceedingly bad description, or no description at all.

Now one such problem is immediately suggested by comparing Newtonian gravitation, general relativity and Maxwell's electromagnetic theory. In Maxwell's electromagnetic theory we have one very important particular kind of solution to the equations, corresponding to electromagnetic radiation. We know of many applications of this solution, in describing both the transmission from a television station, and the transmission of light and radio signals from the stars. All of these solutions of Maxwell's equations have one property in common which is described by saying that they are radiation solutions. This is not an easy matter to explain in non-technical language but it will be sufficient for our purposes if we think of it as simply meaning that they correspond to electromagnetic fields which behave roughly in the manner of radio station transmissions, light etc. The question now arises as to whether there are radiation solutions of the field equations of gravitation. One of the important properties of the radiation solutions of Maxwell's equations is that they travel with the speed of light. In Newtonian gravitation there is no limiting speed like that of light and the gravitational interaction between bodies is propagated instantaneously. If we could imagine the sun suddenly being destroyed, the earth would immediately start to move in a straight line, instead of its elliptical orbit, although it would be 8 minutes before darkness descended on the earth. Thus Newtonian gravitation is an exceedingly poor approximation to a theory in which gravitational radiation, if there is such a thing, plays an important part.

Accordingly the question of the existence of gravitational radiation and its nature were considered very important problems in relativity in the 'fifties. As far as the theoretical side is concerned the nature of gravitational radiation is exceedingly well understood now. Consider two or more bodies that are initially in a stationary state, so that the gravitational field surrounding each of them is roughly the Schwarzschild solution (critical radius). If these bodies then move about relative to each other, finally assuming a stationary state again, they will emit gravitational radiation between the initial and final states. Moreover the final mass of the collected bodies is less than their original total mass by the amount of energy carried away by the radiation. So the theoretical description is perfectly satisfactory, but what remains is to find something about the experimental detection of such waves. A demonstration that such waves exist would certainly mean that much more was known with certainty about general relativity, and this could be of considerable importance in cosmology.

The great authority on the detection of gravitational radiation is Joe Weber in Maryland. The position of Weber might be likened to that of Heinrich Hertz. After Maxwell had published his equations and so predicted the existence of electromagnetic radiation Hertz was able to set up a spark gap and a receiver. By altering the relative positions of the spark gap and the receiver he was able to determine the wavelength of the radiation with which he was dealing. Moreover, most important of all, he was able to show quite clearly that the radiation travelled from the transmitter to the receiver by the simple expedient of switching off the transmitter when the receiver ceased to have a spark in its gap. Unfortunately in the gravitational case Weber is by no means in such a happy position. There are no transmitters of gravitational radiation of anything like sufficient power. Basically this is because the gravitational force is so very much weaker than the electromagnetic force. Any of the bodies which can be moved about in the laboratory would, according to the theory, produce gravitational radiation, but it will be of such a small amount that the possibility of detecting it is very far indeed beyond the limits of experimental feasibility. Accordingly the only possibility is just to sit and wait for gravitational radiation to come from some major catastrophe in the universe. It does not take very long to guess that the sort of catastrophes that have come to mind are probably the inevitable collapse of massive bodies and similar phenomena.

In order to discuss Weber's experiments we must first say a little about gravitational radiation. General relativity is a considerably more complicated theory than Maxwell's electromagnetism. If we want to detect an electromagnetic wave we need some sort of aerial, basically a form of dipole. We want to think of this in the simplest theoretical way possible, without bringing into question properties of steel rods and so on. It can be considered as a pair of charged particles at a certain distance apart, with a wave travelling past them. When the electromagnetic wave passes the separation of the particles alters. The gravitational wave is somewhat more complicated than this. To detect a gravitational wave one needs a system of 4

The aluminium cylinder used by Weber to detect gravitational radiation. Clamped around the centre is the band containing the sensitive crystals

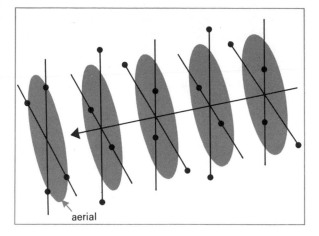

aerial

As a gravitational wave (indicated by the arrow) passes a simple aerial bearing four particles arranged on two diameters, the particles move relative to each other

particles arranged as if on a circle. When the gravitational wave passes through, two of the particles, at opposite ends of a diameter, will move towards each other while the other pair, at the opposite ends of a diameter at right angles to the first one, will move away from each other.

We must have this much technical knowledge to understand the basic construction of Weber's apparatus. He employs an aluminium cylinder about 5 ft long and 2–3 ft in diameter. Round the centre of this cylinder are clamped a number of crystals and these are wired up to a very high gain DC amplifier. When the gravitational wave passes the cylinder it sets up a compression situation in which parts of the cylinder go inwards, other parts go outwards and the crystals correspondingly produce electric currents which can then be amplified by the amplifier. So far the apparatus sounds fairly simple. Now it is necessary to meet the difficulties which are caused by the very low power of the radiation expected. This will involve building a very high gain amplifier indeed. It follows that the resulting apparatus will also be able to detect local disturbances very well. If someone should walk noisily past the apparatus, or a lorry pass on the road outside, a large signal will appear. All of these spurious signals must be eliminated. At first Weber worked on the principle of building a well insulated set of apparatus but not all external noises can be avoided in this way; earthquakes, for example, are liable to overcome even the best insulation. There might also be extreme electromagnetic radiation from the sun and cosmic ray showers, all of which would disturb the electrical part of the apparatus considerably. These difficulties Weber avoided by using not one detector but two, and at a considerable distance apart.

This may not be the only way. Indeed, the reader may have wondered why such a large aluminium cylinder had to be employed. The reason is that the sensitivity of the detector will depend partly on the mass available. Weber's first idea was to use the largest available mass that he had, that is, that of the earth, and this would be much more sensitive than the aluminium bar. However, the earth is much too noisy, since all kinds of earth tremors, both natural and man-made, are quite impossible to eliminate. Another possibility, however, would be to use the moon as the mass and to fix the crystals on it when astronauts visit it. This would perhaps be an effective way of detecting gravitational radiation, but so far it has not been used. There are some hopes that the US Moon programme will be used for this purpose soon.

Aldrin fixing equipment to the moon's surface to relay details of moonquake activity back to the earth. Instruments of a somewhat different kind, preferably fixed at two points fairly widely separated on the moon's surface, could give very valuable information about gravitational radiation, and will probably be transported in the near future

For the present then, in order to avoid all the spurious signals two widely spaced detectors are used, one at Maryland and one at Argonne. The nature of the waves observed by Weber are spikes above a general threshold signal which is just produced by noise. Taking two observations at the two laboratories he defines a coincidence to mean two spikes above the threshold which agree in time to within 0.44 of a second. He considers that such a coincidence could be merely an accidental one about once in several years. But in fact he observes many more coincidences, roughly 2 or 3 per week. By a careful process of elimination he manages to show that nothing else can be causing these except gravitation. He uses the height of the spike as a measure of the strength of the signal. The aperture of the apparatus is about a 70° beam which sweeps across the sky as the earth rotates. Accordingly Weber manages over several hours to get a diagram showing whether the intensity is more in one direction than another. There is considerable indication that there is a maximum of the signal in the direction of the galactic centre.

If this is the case, however, then it seems that it is not a question of collapsing stars, supernovae, or pulsars so much as something in the centre of the galaxy which is causing the radiation. The amount of energy in one of the pulses seems to relate to an amount of energy corresponding to more than the mass of the sun, though here we have to be cautious, for we are very much in the dark about how to do the calculations. Weber originally made the natural assumption that he had not been particularly favoured by providence in his choice of the natural frequency to which his apparatus responded. Thus he expected that the amount of energy observed at the one frequency at which he was measuring would be much the same as at any other frequency, at least over a fair-sized frequency band. If so much energy were to come from collapse it would be from the catastrophic collapse of a body of at least ten times the sun's mass, collapsing in a time of about a thousandth of a second. It is fairly evident

171

that we need a pretty high efficiency of conversion into gravitational waves to produce energies of the sort observed. Moreover if everything is coming from the centre of the galaxy we have to be prepared for our galaxy to be losing a mass of something like 300 times the sun's mass each year, and this is a large amount.

More recently, however, Weber has been able to modify his apparatus so as to receive waves of a different frequency, and it now appears that the waves may occupy a very narrow frequency band. If this is so it means, of course, that he has been very fortunate in his original choice of frequency. But, what is more important, it reduces the total amount of energy that is likely to be needed.

Of course there is always a possibility which tends to be forgotten in the excitement of discoveries about very distant matter. It is just possible that by a fortunate chance we are, in astronomical terms, close to a neutron star which happens to be having earthquakes (or starquakes) two or three times a week. It is thought that neutron stars have a hard crust so that some sort of earthquake situation is quite possible on them. This could have produced the radiation, in which case we would really know very little more about the universe as a whole from it. This possibility, which is an extremely unlikely one, reminds us that general relativity is still relevant, for the suggestion will naturally arise in the reader's mind that such a close neutron star would be optically visible. If we take a neutron star to be one which has contracted so much as to be inside the critical sphere which general relativity predicts for each mass, the answer seems to be that it would not be visible. Suppose that we have a mass m and we calculate its critical length Gm/c^2 and draw a sphere whose radius is equal to that critical length. Then we find that, in the case of the mass being all inside the sphere, the surface which the sphere provides is a very curious one. It will allow signals to enter but it will allow none to leave. It is a kind of one-way surface within the space that it occupies. Any light or radiation produced inside the surface will not be able to get out so the star cannot be seen at all. Such stars, now known as 'black holes', could be detected by their gravitational field quite apart from their giving out gravitational radiation, and although no such star is known it is not impossible for one to exist.

We must not be misled, however, by the importance which general relativity has had in all our considerations into supposing that it is a major part of theoretical physics. It is indeed a minor part, although an important one, and it is a minor part because it is so unconnected with the rest of theorectical physics. There cannot be any wholly satisfactory cosmological theory until we know how to connect general relativity with quantum mechanics. One instance of this was described in the last chapter where a paradox arose in the hybrid theory which takes some ideas from general relativity and some ideas from quantum mechanics. Baryon conservation seems not to hold when a massive body is inside its own Schwarzschild sphere. We have no idea whether a combined theory which included both relativity and quantum mechanics would resolve this paradox or not, but we should certainly hope so. In the meantime, in the absence of such a theory it is nonetheless necessary to look at some of the experimental facts and the theories in the field of modern quantum mechanics because they have some relevance in cosmology.

One of the most puzzling features in the atomic field is the very large number of particles, formerly called elementary particles, which are now known to exist. The situation has changed dramatically in the last ten years. A reasonable classification of these particles has been produced and the most plausible version of this, due to Murray Gell-Mann in 1964, is that which employs the so-called SU_3 symmetry theory. This builds up all the best known of the elementary particles from a single kind of fundamental building block called a quark. The most peculiar characteristic of the quark is its charge. Although all the observed particles have charges which are integral multiples of the electron charge e, the quark has a charge of either $\frac{2}{3} e$ or $-\frac{1}{3} e$. In 1964 and '65 a great deal of effort was devoted to looking for isolated quarks on the earth, but without success. In 1968 J. C. Huang and T. W. Edwards proposed to connect the problem of looking for quarks with the problem of explaining the energy source of quasars. In other words they supposed that it might be possible to observe quarks somewhere other than on the earth at first. There is an analogy here. Helium was first observed on the sun, before it was found on the earth, by the helium lines in the sun's spectrum.

In principle the idea of Huang and Edwards is quite simple. They suppose that some atomic nuclei in the quasar have one or more quarks missing. As may be expected, this has an important effect on the wavelength of the lines in the spectrum of various atoms, which will alter the general pattern of the line spectrum. If an astronomer looks at this altered pattern of lines, it is reasonable to expect that he will make some false identifications, because the lines will be in different places relative to each other from those which he had expected. By a detailed analysis of the elements which one would expect to find and a comparison of the lines from the quasar 3C191 Huang and Edwards reckoned that the very large red-shift, in this case 1.95, was not entirely a cosmological one. In fact they estimate that the genuine red-shift is only 0.31, the rest being a spurious shift due to the change in the wavelength of the lines by the missing quarks. This new value for the red-shift, bringing the quasar much closer, would correspond to an energy output rather less than that of the brightest known galaxies. Naturally there is a great deal more work to be done before any such identification could be accepted with any degree of assurance. But the very idea of such an observation raises in a most interesting form the mutual influence of general relativity and quantum mechanics on each other. This influence has many different aspects. For example, the whole of the explanation of the internal constitution of the stars and the way in which their light and radio emission is produced requires a quantum mechanical theory. Accordingly the actual means of observation in cosmology depend heavily on the existence of quantum mechanics. On the other hand the collapse mechanism for the production of the quasar energies, if indeed this is responsible for the production of these energies, seems to influence quantum mechanics since it predicts the destruction of baryons. We feel the need for a comprehensive theory which will contain both relativity and quantum mechanics in such a way that there is no inconsistency between them. Exactly what form this would take is still a mystery, however. The next chapter considers some unorthodox views of the union.

UNUSUAL COSMOLOGICAL THEORIES

Whilst the rival hosts are drawn up on either side to defend the embattled positions of the evolutionary cosmology and the steady-state theory there are still many minor skirmishes taking place in the wings. This chapter is intended to deal with some of those. We may start with the question raised at the end of the last chapter about the union of general relativity and quantum mechanics. Apart from its obvious value to cosmology, there is a general reason for looking at the relationship between general relativity and quantum mechanics. It is hardly possible to hope for an adequate explanation of Mach's principle without some such comprehensive theory. For, after all, Mach's principle states that phenomena on the very large scale influence local phenomena. In other words, an adequate description of Mach's principle would also contain a description of the relative influence of gravitational theory and quantum mechanics.

One attempt to unite these two theories is known as quantizing the gravitational field. Quantum mechanics arises by a process which is somewhat mysterious even to its practitioners. This process is known as quantization. Essentially, one takes a theory of physics which is perfectly continuous in all its predictions, and one makes alterations to it in a way which has been successfully done with other theories. These alterations give some of the predictions of the theory a discrete

Opposite: The night sky still moves us to ask fundamental questions, as it has done throughout history. Is the universe evolving or is it the same from an infinite past to an infinite future? Or something different from either? We continue to seek a universal theory, and observational evidence – and so will our children and our grandchildren

value, and so integers now enter the theory. The way in which Schrödinger thought about this originally was in connection with a stretched string. A string of infinite length can have waves passing along it of any wavelength. If a string is fixed at both ends, however, the wavelength of standing waves in the string is determined by the string's length, which is of course the principle of all stringed musical instruments. The general continuous theory of the infinite string is therefore modified by the string being fixed at the ends, and as a result certain discrete quantities become involved. These are the frequency of the fundamental mode and all those multiples of it (harmonics) which are also excited.

In general it could be said that the process of quantization amounts to this: one takes a classical theory and one performs an analysis of the observations which it predicts in terms of pendulum-like motions. This tradition really goes back to the moment when Galileo timed the swinging of the heavy lamp in the cathedral at Pisa with the help of his pulse. The pendulum is regarded as the fundamental model in terms of which everything can be understood. When the analysis has been carried out in terms of pendulums the rest is easy, because for the case of the pendulum (or what is known technically as the harmonic oscillator) the quantization process is completely understood from experience, and has been carried out in a number of different ways since quantum mechanics began.

Such a process works in a certain sense for field theories like the electromagnetic field in which there is a principle of superposition. If one asks for the force on a certain charge A due to another charge B and also for the force on A due to a third charge C, then the force on it due to the two charges B and C taken together is the sum of the individual forces. The same is true for masses in Newton's theory of gravitation but an essential feature of general relativity is that when two masses act on a third mass their effect is not the same as the sum of their separate effects, but there is something additional besides. This is what is meant by saying that electromagnetism is a linear theory but general relativity is a non-linear theory. Quantization is a process which has been devised especially for linear theories and so it is a matter of doubt whether the same process would be applicable for a non-linear theory.

There is however an even more serious difficulty in applying the process to general relativity. The so-called gravitational field in general relativity, as has been said before, is replaced by being made part of the background geometry of space. There is no room left for rigid fields of the kind which occur in quantum mechanics, or for the complementary representation of fields in terms of particles and particles in terms of fields. The only way out would be to find some alternative representation to the geometry of a particle-like form and we have no idea of what this would be. There has nonetheless been an extensive programme to quantize general relativity over the past 20 years, but it has not yet been successfully done.

Three radical theories

It is therefore worth asking whether there is some different way of forming a union of the theories. Two different ideas have been put forward, which will be dealt with in turn. The first is due to P. A. M. Dirac who noticed that most of the fundamental physical constants have to be expressed in specific units such as centimetres or seconds or grams. But it is possible to construct from them certain new constants which have the same value in all systems of units. These are the dimensionless constants. Dirac took the rather unusual view that only the naturally occurring numbers of this kind should be considered. He allowed himself to prepare his experimental material, as it were, by replacing any of these numbers which were less than one by its reciprocal. The result was a set of numbers all greater than unity. He then found that these numbers 'clustered'.

For example, the rate at which the galaxies expand, that is Hubble's constant, can be expressed equally well in terms of another constant of the nature of a time. This constant is often called the age of the universe, meaning by this the age which the universe would have if it had expanded at this constant rate since it was first created. (It is actually the reciprocal of Hubble's constant.) At the time of Dirac's first paper, this age was estimated to be about 2,000 million years, but it would now be put rather higher than that. This has been given in terms of years, but there is also a unit of time in atomic physics which is defined by atomic constants. If e is the charge on an electron, m its mass and c the velocity of light, then the quantity e^2/mc^3 is actually a very short time. Thus when the age of the universe is expressed in terms of this unit, a very large number is produced. Since this number is the ratio of two times, its value does not depend upon the units used to measure it. If feet and seconds were used instead of centimetres and years the same answer – a figure of about 10^{40}, or one with 40 noughts after it – would result. Dirac was also struck by the fact that gravitational force is also about 10^{40} times weaker than the electrostatic force. Furthermore, cosmologists have estimated that there must be a total of about 10^{80} particles in the universe – and this number, of course, is the square of 10^{40}. It really is very surprising indeed that these three simple measures of the universe – its age, the number of particles it contains and the ratio of two of its most significant forces – should be so simply related. But Dirac was intrigued by some even more fundamental facts.

According to Dirac all the naturally occurring dimensionless quantities fall into one of exactly three classes; they are either of the order of one or of order of the age of the universe in atomic units or of the square of this. To explain what is meant by 'order' it should be said that those of the order of one, for example, vary between one and perhaps a couple of thousand. There is a similar spread with the constants in the other two groups, but the groups are so very widely removed from each other that there is no possibility of any confusion. The numerical coincidences which Dirac had

P. A. M. Dirac, most famous for his discovery of the equation describing the electron in a relativistic manner

noticed here had been spotted by earlier workers. For example Stewart, in 1931, had published a note in the *Physical Review* on a different coincidence from the one which has been described. But Dirac was the first person to propose a theory to explain them, namely that the ratios in question, that is the age of the universe and the number of particles, were related because the physical constants which occurred in them, such as the gravitational constant, depended upon the time that had elapsed since the creation of the universe. The fact that one of the constants was the age of the universe was significant for Dirac. The others were then either equal to it, or to its square, because they depended upon this age; the constants in the first group, which were of order one, were grouped together because they did not depend upon the age at all.

It is not fair to call Dirac's theory a complete cosmology, though he did elaborate it to some extent in 1937. He takes a rather similar model of the universe to the usual expanding models of general relativity, but he makes his postulate about the dependence of constants on the age of the universe more specific in the form: any two of the very large dimensionless constants are connected by a simple equation, whose coefficients are of order one. It is possible, by a very simple argument about the average density of matter and Hubble's constant (which are of the same order) to derive the dependence of Hubble's constant on time. It then turns out that the actual age of the universe is only one third of what we would expect by using the present value of Hubble's constant in the normal expanding universe models. This shortening of the age is sufficient to lead to a contradiction with the ages of the stars, but Dirac believes it possible to dodge this by assuming that nuclear processes occurred more rapidly compared with atomic processes in the past than now.

It is now recognized that Dirac's theory suffers from some very severe mathematical problems. But it has provided at least one lasting insight. It frequently happens that the quantities whose ratios we have been discussing are respectively, one cosmological and one atomic quantity. It is possible to form dimensionless ratios in atomic physics alone or in cosmology alone, but these do not seem to be of such great importance. Most of the dimensionless ratios which occur mix atomic physics and cosmology and accordingly a ghost of Mach's principle is seen again. Most people agree that the numbers which occur must be significant in some way and if so then we have again a connection between the very large and the very small of the kind that Mach pointed out.

Jordan's theory, based on Dirac's argument, reached a finished form by 1947. He avoids some of the difficulties by abandoning the law that the total mass of the universe is constant. In this he is at one with the steady-state theorists, but his reasons are quite different. He believes (erroneously) that in order to avoid Olbers' paradox one must have a curved space. In fact curvature does not affect Olbers' paradox at all. Jordan then proposes to keep the total energy constant by supposing that the new mass is created in an extremely condensed form and then explodes. In this way, the increase of energy caused by the increase of mass is exactly counterbalanced by the negative gravitational potential energy of the condensation. Jordan believes the explosion to result in a supernova explosion; and calculates from his theory that there should be about one such explosion per galaxy per year. This is an extremely precise prediction, but since the observed rate of supernova explosions is one per galaxy per two or three centuries the theory is in hopeless conflict with observation.

No mention of the dimensionless constants would be complete without a reference to Eddington. Nowadays his writings attract much less interest than they did – indeed, less than they warrant – but if we are to give anything like an adequate survey of unusual cosmologies, his must play a part. His theory is not primarily a cosmological one, since it is intended as a scheme for physics as a whole, but in view of Mach's principle it is bound to have cosmological significance.

Eddington's work as an astronomer brought him early into close contact with general relativity and he was instrumental in popularizing this with English scientists by the publication of his book, *The Mathematical Theory of Relativity* in 1924. This was the first readable account of the theory in English. Already in the later parts of this book he begins to show a slightly different outlook on scientific theories and their relationship with observation from the orthodox one. He speaks of the process of theory construction in general relativity, and then of facing the theory with experiment, as a process in which the theory is freely constructed and then the experiment is used simply as a means of identifying the quantities which enter into the theory.

Shortly after the publication of this book, Dirac showed how the equation which describes the electron

Sir Arthur Eddington

in quantum mechanics could be written in a form which was completely consistent with special relativity. This was a most satisfactory state of affairs but for Eddington it was a tremendous shock, because the mathematical techniques used in re-writing the equation were not of the kind which he had been developing. In fact, an entirely new mathematical technique was employed by Dirac. It is impossible to overestimate the psychological effect of this discovery on Eddington; he mentions it over and over again as the starting point of his investigations. He decided to elaborate this technique in order to get to the real explanation of how the mathematics had managed to throw up an equation which he himself had never suspected.

He soon found that the elaboration of Dirac's technique led to highly significant numbers. Amongst these the most notorious has been the ratio of the masses of the proton to the electron, a quantity which is measured to be about 1,836, and which Eddington found by a fairly simple theoretical argument to have the value 1,848. Another was the so-called fine structure constant which occurs in the expression for the fine structure of the lines in the hydrogen spectrum, which Eddington calculated to have the value 137, and whose observed value (137.036) is not far from this.

Two unfortunate developments now arose. In the first place, the calculation of numbers of this kind, which were normally thought of by physicists as experimental results, apparently from an *a priori* position led to intense criticism. Initially this had the good effect of making Eddington try to deepen the physical foundations of his ideas. But the continuing attack eventually turned him away from listening to all criticism and the last ten years of his life from 1935–45 were spent in a rather lonely way working out his ideas to his own satisfaction. Secondly, the criticism of the disagreements between the calculated values of the constants and their observed values pushed Eddington into an untenable position. He calculated enormous numbers of constants and insisted that his calculated values were exactly right, so long as the experiments had their results reduced in the proper way. This had the effect of diminishing still further the confidence of orthodox physicists in his methods.

As a cosmological theory, Eddington's is quite conservative. He always supposes that the universe is one of the expanding models of general relativity, with the additional restriction that it has expanded from the initial Einstein state. But this lack of adventure in the details of the model is compensated by the boldness of his attempt to construct a new kind of theory.

Today, it is by no means obvious that a valid theory of the kind conceived by Eddington can in fact be constructed. But, again, a careful study of Eddington's flawed efforts does point to something new. At the moment this presents itself as a difficulty. This difficulty is that the theory needs to work in terms of two theoretical structures, of different complexities, at the same time. It is as if we had to solve problems working partly in Newtonian mechanics and partly in special relativity, and it was no longer possible either to say that all the velocities were small so that Newtonian mechanics would be sufficiently accurate nor could the problem be completely formulated inside the more complex theoretical structure of special relativity.

This directs attention to a problem whose solution would really contribute to cosmological progress. This problem is to determine what is the relationship between two theoretical structures one more complex than another, or, to be a little less ambitious, to determine what relationships there are between numerical quantities associated with such structures. If we could understand this problem, then we could understand how general relativity and quantum mechanics would be related and so we might then hope to find an adequate description of the universe as a whole. It is scarcely possible to hope that such a description could be found in any simpler way. In some ways this is a counsel of despair; in other ways it is an optimistic look to the future.

The difficulties in cosmology which have already been pointed out by the steady-state theorists are even larger than they had supposed. The argument in favour of the steady-state theory was that, if it is not true, then cosmology is a very much more difficult subject, since we do not know how physical laws change from one part of the universe to another. There is no difficulty in re-phrasing this in terms of the abstract structures of this chapter but now the difficulty in a comprehensive theory of cosmology is seen to be more complicated than whether we are to make the steady-state assumption or not. Indeed, the plausibility of making this assumption has in recent years been considerably reduced by the experimental results, although one can still hope that these will be altered. Rather it is whether the observational results can be seen as part of one comprehensive theory or whether we need more than one. It now seems astonishingly naïve to suppose that there could be one comprehensive theory which would collect together all the experimental results of different kinds. But if more than one is needed, then the question of relationships between theories which cannot be considered as parts of a general comprehensive theory is a very difficult one indeed.

The theories of Dirac and Eddington have this in common: they make a really radical break with orthodox physics in one way or another. To redress the balance we shall conclude with two theories that strive to explain as much as possible within the orthodox framework of physical theory, making, instead, rather radical changes of view about the nature of the matter in the universe.

The bonds of orthodox physics

The first of these seems to have come to nothing, so we shall mention it only briefly, but it is worth including because it shows how careful one needs to be in questioning one's assumptions. This is the so-called 'electric universe' of R. A. Lyttleton and Bondi, which they formulated in 1960. Lyttleton, in particular, was concerned over the question of the expansion in the universe. Why, he wondered, in a mass of gravitating matter, which would therefore be expected to collapse, was there a predominant motion of recession? It was almost as if some repulsive force was acting. In search of a repulsive force, one's mind naturally turns to the electric force between charges of the same sign.

Now there is undoubtedly a great deal of electric charge in the universe since matter is made up as we know of protons, electrons and neutrons. Only the last of these is a neutral particle, but the effect of the charge on the other particles is completely neutralized because the charge on the proton is positive and of the same size as the negative one on the electron.

Moreover there is a great deal of evidence in quantum mechanics that any two electrons are exactly alike. It is a curious feature of all the fundamental

Dr R. A. Lyttleton

building bricks of matter that any two of the same kind have this curious exact similarity. It is sometimes expressed by saying that fundamental particles are characterized by having a finite number of attributes. Of course there is a sense in which one can say that any two cricket-balls are also alike. But we know that if we take two balls which any two teams are certain to accept as equally appropriate for their game, we could, by a sufficiently refined analysis, find a small difference. They have the same mass, up to a certain limit of accuracy; but that is not to say that, if we make very accurate mass measurements, we shall not find a difference of a small fraction of a milligram. Similarly there may be minute differences in hardness, colour and so on.

But the case with electrons (or other elementary particles) is quite different. By comparing the masses or charges of electrons we can, indeed, draw only the same approximate conclusions as with cricket balls.

But we also have, in quantum mechanics, quite different methods of experiment. Because two electrons are exactly alike, a system is wholly unchanged if the two electrons are interchanged. This gives rise in suitable circumstances to an additional energy 'exchange energy' making a significant contribution to quantum interactions. But if the electrons are only approximately the same, the exchange possibility vanishes.

Now it was thought by many people that there were also arguments to show that the magnitude of the proton charge was exactly equal to that of the electron charge. But Lyttleton noticed that no such evidence was available; the evidence in this case was just of the same kind as with cricket-balls, i.e. subject to experimental error. Accordingly he and Bondi argued that in fact the charge on the proton was not exactly equal and opposite to that on the electron. It follows that in a piece of so-called 'neutral' matter there will be a certain residual charge, arising because the electron charges are not exactly balanced by the proton charges. Such an excess of sign of one charge will cause a repulsion, and Lyttleton and Bondi attributed the expansion of the universe to this repulsive force.

It turned out, rather surprisingly, that the amount of excess of one charge over the other, in order to provide the measured expansion of the universe as well as to account for the observed cosmic ray phenomena, was just slightly less than would have been noticeable in the laboratory measurements. This makes the theory a striking example of how recognition of one single unusual assumption can lead to a remarkably changed theory. But since 1961 more careful measurements of any excess of charge on either proton or electron have been made both in order to test the theory and out of general interest. It now seems as if the measured excess is definitely much too small to let the Lyttleton-Bondi theory work.

The remaining theory is much more ambitious than that of Lyttleton and Bondi. This theory was originally due to Oskar Klein but has been much developed by Hannes Alfvén. Whether or not it holds the key to the solution of all cosmological problems, as Klein and Alfvén seem to think, is still a matter of some doubt, but the theory is of the greatest interest in showing that it is perfectly possible to construct inside a very orthodox framework a theory which is significantly different from either the big-bang or the steady-state theories.

The basis of the theory is really the whole-hearted embracing of the idea of anti-matter. It has for many years been commonplace in quantum mechanics that to every particle there corresponds what is known as its anti-particle. If the original particle is charged the anti-particle has the opposite sign of charge. When a particle and anti-particle are brought together they annihilate each other with a subsequent enormous release of energy. As far as we can see, the laws of quantum mechanics are completely symmetrical with regard to matter and anti-matter. On the other hand the world in which we live is undoubtedly predominantly composed of matter.

It is an exceedingly difficult experiment to produce anti-matter locally, although there have been successes, as in the production of the positron, which is the

The plate enabling Anderson to discover the positron. Anderson studied many ·such photographs of positively charged particles in cloud chambers, trying to reconcile the tracks with the particles being protons. As it became clear that the particles possessed a mass very little different from electrons (protons have a much larger mass), he realized they must be anti-electrons, or positrons

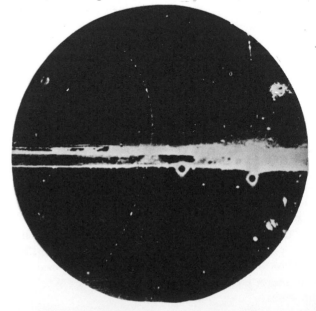

anti-electron, as long ago as 1930 by C. D. Anderson. The anti-proton has also been produced in the past couple of decades. The experiments are so difficult because it is necessary to produce such a particle well away from any ordinary matter, since as soon as it encounters ordinary matter it is annihilated, with the release of energy mentioned before. We have then a paradox in that the physical laws are symmetrical for matter and anti-matter, but our universe, if we go by what we see in our immediate neighbourhood, is completely unsymmetrical.

Before we consider Klein and Alfvén's arguments about the way that anti-matter enters we ought perhaps to say something about their views on the expansion of the universe, irrespective of how this expansion is caused. It is usually considered that, since all the distant nebulae are receding from us with a speed proportional to their distance, there must have been an original singular event of creation (unless we seek a way out by means of the steady-state hypothesis). However this argument is not a realistic one. One measures a number of nebulae at certain distances with certain speeds. Naturally the speeds are not *exactly* proportional to the distance, and if we work out when the nebulae would have been near us (corresponding to the singular event) we would not get exactly the same value for every nebula. In fact, putting in the sort of figures one can deduce from observation, it is quite possible that two different nebulae would actually have been in our local region at periods some 10 million years apart. They need never, then, have been near each other, and there need never have been a singular event. In short, while it is true that from a singular event it is easy to deduce Hubble's law the reverse derivation is impossible. As one example of an alternative explanation of Hubble's law, one can imagine an immense mass of gas which begins to contract under its own gravitational attraction. As the contraction proceeds local condensations are formed, which would be nebulae, and these nebulae move relative to each other in hyperbolic orbits, eventually

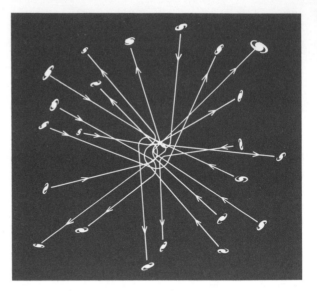

Professor Hannes Alfvén, who developed Oskar Klein's cosmological ideas into a theory (right) that necessitates neither a big bang nor a steady state. An initially diffuse cloud of matter falls together, moving round the centre of mass in hyperbolic orbits, and then diffuses again. We happen to be witnessing the expanding phase

moving round each other and out to a great distance again. Irrespective then of the question of the symmetry of matter and anti-matter, Klein and Alfvén would not believe in the necessity of either a big-bang or a steady-state theory.

Let us turn now to the question of whether it is at all easy to tell if particular nebulae or indeed particular stars in our galaxy might perhaps be made of matter or anti-matter. In view of the complete symmetry between particles and anti-particles most of the physical properties of anti-matter will be exactly like those of ordinary matter. For example, when a large number of anti-hydrogen atoms (that is atoms consisting of a central anti-proton with an anti-electron in an orbit round it), combined with an appropriate number of anti-oxygen atoms to give anti-water we can be sure that this anti-water will freeze at $0°C$, boil at $100°$, make snow flakes and so on. We shall only notice any difference when it comes in contact with ordinary water and we see the resultant explosion. So we can have anti-stars giving out radiation of very much the same kind as stars. We have no way of distinguishing stars from anti-stars by their radiation.

A little more needs to be said however. The space beyond the stars is not completely empty; there is a thin plasma which falls in from the surrounding space into some stars and is ejected from others. If there were anti-stars their plasma would consist of anti-matter. There is therefore some possibility of an interaction here. How much of the world, then, can we be sure consists of matter? Certainly we know that the earth and the moon do. The sun emits a plasma which reaches the earth and causes the aurora borealis. If the plasma were of anti-matter the aurora would be about a thousand times its present brightness. Moreover this plasma reaches Mercury, Venus and Mars, and since we do not see any obvious annihilation phenomena, these planets must also be ordinary matter. Once we reach the outer planets however we have really no evidence at all.

When the solar plasma – ionized gas ejected from the sun – reaches the region of the earth and interacts with the thin topmost layers of the atmosphere, the aurora is formed. If the sun were anti-matter, the aurora would be much brighter because of the mutual annihilation of matter and anti-matter

How much of the universe consists of matter – basically electrons and protons – and how much of matter's mirror image, anti-matter, built up of positrons and anti-protons? Apart from the sun, the moon and the nearest planets, there is no way of telling. It is reasonable to suppose that the universe as a whole is symmetrical, and it could even be that the observable nebulae are alternately matter and anti-matter – as is symbolized by this positive-negative pair representation of the Whirlpool nebula

What we can expect is that if some regions of the universe contain anti-matter and others contain matter there will be somewhere between them a region where the continuous plasma from one interacts with that from another, with consequent radiation. This could be connected with cosmic rays. With the conditions which one would expect to find in space, about half of the energy released in this way could be expected to be in the form of neutrinos, about a third is gamma rays and about a sixth radio waves. Evidently the methods of observation which we have discussed in earlier chapters are going to be extremely important in deciding whether the amount of these signals is con-

sistent with annihilation of the kind suggested. As we said before, neutrinos are very hard to detect and gamma rays likewise, and for this reason the radio waves, although the rarest, are the most likely source of evidence.

Now we come to the question of the symmetry of the universe. We can satisfy the demand for symmetry in a number of different ways. Either we can suppose that the anti-matter is present in some very remote region, and more or less everything that we observe consists of matter, or we may suppose that the observable universe is symmetrical, and this might arise by every other nebula being made up of anti-matter.

On the other hand the symmetry might arise with a symmetrical situation inside our own galaxy, and in every nebula, and here again the choice is either that the remote part of our galaxy consists of anti-matter or that every other star in our vicinity (and elsewhere) consists of anti-matter. Even this extreme kind of symmetry cannot be ruled out by observation at the present time. If Sirius, the brightest star that we see, consisted of anti-matter we should have no means at all of discovering this fact or refuting the suggestion if it were made.

It is now time to put together the two separate discussions which we have given of Klein and Alfvén's views. They based their cosmology on two principles. (1) No new natural laws are to be assumed, (2) there is symmetry between matter and anti-matter. Since there are to be no new natural laws they feel that an original singular creation event is impossible to discuss and they therefore begin with an exceedingly sparse cloud of gas which condenses, forming nebulae, moving under their gravitational attraction, and then ultimately receding. However, a more detailed analysis of such a process on the basis of symmetry between matter and anti-matter would lead to a large amount of annihilation. Ultimately the whole mass would be annihilated. Something must intervene to save the situation and this something is the large amount of radiation pressure set up by the radiation emitted in the annihilation process. This radiation pressure has the effect of causing the matter to blow apart in a more explosive way than it would otherwise.

Naturally the analysis is very difficult and has only been carried out approximately, but Klein and Alfvén are optimistic about the way in which it fits observed phenomena, in particular the speed of recession, cosmic ray observations, radiation, neutrino emission and so on. Whatever may happen to such a theory in the future – whether later observations show it to be completely untenable or justify it absolutely – it stands as a most striking example of an alternative to the two established ways of thought in cosmology. To choose between it and any of the other theories described earlier in the book we need more and more observation. In the process of these observations we shall hope to learn a great deal more about the structure and behaviour of the universe as a whole.

GLOSSARY

acceleration: the rate of change, in magnitude or direction, of the speed of a body. Newton's laws state that acceleration is proportional to the force on a body, the constant of proportionality being the mass. The statement that force = mass × acceleration holds true only to a set of observers who are moving uniformly relative to each other and define the set of inertial frames of reference. These mechanical considerations are connected with the rest of physics by rephrasing them in terms of **energy**. Small modifications in this scheme are necessary when the curious properties of the speed of light are taken into account, giving rise to **special relativity,** and when energy comes in small parcels, or quanta, not continuously. In this case we speak of the quantization of a theory, or of the quantum theory.

arrow of time: the direction of progress of events that accounts for the irreversibility of some physical laws, e.g. the spreading out of waves on water rather than their convergence on a point of disturbance.

baryon: see **elementary particle**

big bang: the name given to the evolutionary theory of the origins of the universe, and also the name describing the initial conditions of the evolutionary model of the universe. The amount of matter in the universe remains constant in this theory, although initially it is in a state of extremely high compression.

black-body radiation: thermal **radiation** reflected round the inside of a closed container with which it is in thermal equilibrium – in other words it is neither transmitting nor absorbing radiation. This is the theoretical situation which led Max Planck to postulate the **quantum theory.**

black hole: a star that has collapsed beyond the condition of a **neutron star** so that all its mass falls within the **critical radius**. Its presence can only be detected by the strong gravitational field that it gives rise to and which allows radiation to escape.

cepheid: a star of varying brightness whose absolute brightness can be determined by the period of variation. Comparison of absolute brightness, thus determined, with apparent ·brightness makes Cepheids valuable 'markers' for astronomers when calculating the distances of stars.

cosmological principle: the name given to the fundamental starting-point in cosmology, that any point in the universe is much the same as any other (averaged over a sufficiently extensive region). This principle is extended by the **steady-state** cosmologists to the perfect cosmological principle, that in addition any time is much the same as any other time. These principles are obviously connected with the **arrow of time.** Whereas the cosmological principle has no trouble in correlating this in a general way with expansion of the universe from an initial singular state, the perfect cosmological principle must use the arrow of time to show that continual creation and not annihilation is taking place.

critical radius: Einstein's general theory of relativity predicts that if all the mass of a body is contained within a sphere of less than a certain radius (also known as a Schwarzschild sphere), no light or radio signals can escape from its gravitational field. This critical radius varies as the mass of the body. See also **black hole.**

Doppler effect: change of frequency of either sound or electromagnetic waves, observed in an approaching or receding source. Just as the whistle of an approaching locomotive seems to drop in pitch as it passes an observer, so the light from a receding galaxy drops in frequency. See **red-shift.**

ecliptic: the name given to the circular path of the sun relative to the earth, against the background of the stars. (Of course, the stars are invisible when the sun is up, but one can estimate an approximate position for the sun just after sunset or before sunrise, and by doing this day by day the ecliptic was plotted some thousand years or more BC.)

electron: see **elementary particle**

elementary particle: now a somewhat vague term referring to the various 'building bricks' from which matter is constructed, and to the products that arise when such particles interact. In the early years of this century the hydrogen atom was found to consist of a heavy nucleus with positive charge, the proton, and a light negatively-charged particle, the electron, in orbit round it. For more complex atoms it was found that uncharged particles resembling the proton were also present in the nucleus; these were neutrons. Dirac's theory, which made all this consistent with both **quantum theory** and **special relativity,** predicted two new features: to every particle corresponds another one, of the same mass but of opposite electric charge (e.g. to the electron corresponds the positron) and, further, a sufficient supply of energy enables an electron-positron to be created, pair-creation. Special relativity also allows the existence of particles of zero mass, and the most important of these is the neutrino. When these particles interact many short-lived ones are produced, and the large number of these throws doubt on whether or not they are elementary; some are light, like the electron, others heavier than the proton, and these latter are called baryons. Since they carry nearly all the mass, the number of them is unchanged in any reaction. Since these supposedly elementary particles may not be the basic ones (any more than atoms proved to be), theories of their construction from a few even more primitive entities, quarks, have been elaborated, but so far an isolated quark has not been found.

energy: a physical quantity that expresses work done by a body in virtue of its position (potential energy) or its motion (kinetic energy). A car standing on a hill, in neutral and with the brakes on, has potential energy. When the brakes are released it has kinetic energy, which increases with its increasing speed.

galaxy: an immensely distant collection of stars, generically similar to but independent of our own galaxy (the Milky Way). There are millions of galaxies in the observable universe, and the more distant ones have the spectra of their **radiation** shifted towards the red end of the spectrum by an amount which has been found to be proportional to their distance. See **red-shift**.

general relativity: see **relativity**

gravitation, gravitational field: Newtonian gravitation was originally expressed in terms of action-at-a-distance (i.e. stating the force between two bodies), but it is more convenient to express it in terms of one body finding itself in a field of force determined by the other, and this field is referred to as the gravitational field.

inertial frame of reference: see **acceleration**

kinetic energy: see **energy**

light-year: the distance travelled in one year by light, which has a velocity of 186,000 miles per second. It is approximately equivalent to six million million miles (6×10^{12}).

Mach's principle: the determination of local **inertial frames of reference** by the distribution of distant matter.

mass: a force applied to a body produces an acceleration proportional to the force. The constant of proportionality is the mass of the body.

nebula: a cloud of diffuse gas within the Milky Way, or – outside the Galaxy – a separate star system at great distance.

neutrino: see **elementary particle**

neutron: see **elementary particle**

neutron star: see **stars**

pair-creation: see **elementary particle**

particle accelerator: a machine constructed to speed up a stream of **elementary particles** (e.g. protons) to speeds near to that of light and then to direct the beam on a target, in order to observe (on a photographic plate) the resultant reaction between particles. The speeding-up is most commonly done on a large circular track, by means of electromagnetic fields.

planet: satellite of a **star**, particularly of the sun. The so-called Morning Star and Evening Star are both in fact a planet (Venus), i.e. a lump of cold matter reflecting the sun's rays as does the moon. Before a body can shine by itself it must be massive enough (about one hundred times the mass of Jupiter or one eighth the mass of the sun).

positron: see **elementary particle**

potential energy: see **energy**

proton: see **elementary particle**

pulsar: an extremely small body, emitting very con-cise **radiation** at a high frequency and precisely regular intervals of one second or less. Most pulsars, which are a recently discovered phenomenon, are within the Galaxy.

quantization, quantum theory: see **acceleration**

quark: see **elementary particle**

quasar: short for quasi-stellar radio source. A recently discovered class of celestial body, characterized by a **red-shift** so high that it is thought to indicate that they are at enormous distances.

radiation: the collective name for all transmissions across empty space of signals with the same speed as the most common kind viz. light. Radio signals and X-rays are also very important and differ from visible light only in wavelength. The spectrum of **radiation** is the amount of each wavelength it contains, the visible band stretching through the colours from red (high wavelength) to violet (low wavelength). Infra-red and ultra-violet are names given to the radiation immediately beyond the visible range in each direction. In addition to electromagnetic radiation (described here), recent experiments indicate the existence of gravitational radiation (see p. 166 ff.).

red-shift: the shift of the spectral lines of a star or nebula towards the red end of the **spectrum** due to the Doppler effect (when the object is receding from the observer).

relativity: Einstein's Special Theory of Relativity (1905) discussed the behaviour of light, energy and matter in the special case of observers moving at a con-stant speed, and in a straight line, in relation to each other. The case of one observer accelerating with respect to another was covered by the General Theory of Relativity (1916), which formulated gravitation in such a way as to incorporate the Special Theory.

Schwarzschild sphere: see **critical radius**

special relativity: see **relativity**.

spectrum: see **radiation**

star: A vast mass of gas with a central temperature high enough to promote nuclear processes which result in the radiation of various forms of energy. Even before the precise mechanism of this reaction was known, Eddington proposed a theoretical relation between the mass of the star and its luminosity and stars of different masses were observed to have different characteristics. But one type, the white dwarfs, do not obey the mass–luminosity relation. There seems little doubt that these are stars at the end of their life, but whether their luminosity is produced by remnants of hydrogen left in their external regions is now questioned. Instead, a new mechanism of gravitational contraction is now in vogue, the gravitational potential energy being 'tapped' as the star collapses. Whether or not white dwarfs are of this type, such a process must occur, and the high interior densities cause the electrons there to combine with protons to form neutrons. In this way a neutron star can be formed.

steady-state cosmology: the model of the universe, described by Bondi, Gold and Hoyle, which supposes that the universe on a large scale has always been much the same as it is now, the expansion being provided for by the continual creation of new matter.

white-dwarf: see **star**

BIBLIOGRAPHY

General and non-technical books

Andrade, E. N. de C., *Sir Isaac Newton*. London 1954, New York 1958

Armitage, A., *Copernicus*. London 1953

Bok, B. J., *The Astronomer's Universe*. London, New York 1958

Bondi, H., *The Universe at Large*. New York 1960, London 1961

Bonner, W. B., *The Mystery of the Expanding Universe*. London 1905

Butterfield, H., *Origins of Modern Science*. New York 1965

de Santillana, G., *The Crime of Galileo*. Chicago 1955, London 1961

Gade, J. A., *The Life and Times of Tycho Brahe*. New York 1947

Gamow, G., *The Creation of the Universe*. New York 1961

Geymonat, L., *Galileo Galilei*. New York 1965

Hodge, P. W., *Concepts of the Universe*. New York 1969

Hoyle, F., *The Nature of the Universe*. Oxford, Columbia 1960

Hubble, E., *The Observational Approach to Cosmology*. Oxford 1937

— *The Realm of the Nebulae*. London, New York 1936

Jastrow, R., *Red Giants and White Dwarfs*. New York 1967

Koestler, A., *The Sleepwalkers*. New York 1963, London 1968

— *The Watershed*: A Biography of Johannes Kepler. New York 1960, London 1961

Koyre, A., *From the Closed World to the Infinite Universe*. Baltimore 1969

Lovell, B., *Our Present Knowledge of the Universe*. Manchester 1967, Cambridge 1968

Murchie, G., *Music of the Spheres*. Boston 1961, London 1968

North, J. D., *The Measure of the Universe*. Oxford 1965

Pannekoek, A., *A History of Astronomy*. London, New York 1961

Russell, B. (Ed. F. Pirani), *ABC of Relativity*. Revised ed. London 1968

Schilpp, P. A. (Ed.), *Albert Einstein, Philosopher Scientist*. New York 1951

Sciama, D.W., *The Unity of the Universe*. London 1959

Science Year. Chicago (published annually).

Seeger, R. J., *Galileo Galilei: his life and works*. Oxford 1966

Shapley, H., *View from a Distant Star*. New York 1963

Smart, W. M., *The Riddle of the Universe*. London, New York 1968

Velikovsky, I., *Worlds in Collision*. London, New York 1950

Whitrow, G. J., *The Structure and Evolution of the Universe*. London 1959, New York 1961

More advanced books

Bohm, D., *Special Theory of Relativity*. New York 1965

Bondi, H., *Cosmology* (2nd ed.). Cambridge 1960

Einstein, A., *The Meaning of Relativity*. London, Princeton 1956

Herivel, J. W., *The Background to Newton's Principia*. New York 1965, Oxford 1966

Hoyle, F., *Galaxies, Nuclei and Quasars*. London, New York 1965

Kuhn, T., *The Copernican Revolution*. Harvard, Oxford 1957

Rosser, W. G. V., *Relativity and High Energy Physics*. London 1969

Schatzman, E., *The Origin and Evolution of the Universe*. London, New York 1966

Struve, O., Lynds, B. and Pillans, H., *Elementary Astronomy*. Oxford 1959

Wheeler, J. A., *Geometrodynamics*. New York 1962

Wyatt, S. P., *Principles of Astronomy*. Rockleigh, New Jersey 1964

Journals

Science
Scientific American

LIST AND SOURCES OF ILLUSTRATIONS

The diagrams were drawn by John Stokes

INDEX *Numbers in italics refer to illustrations*